THE COMPLETE HOMESTEAD PLANNER

A Month-by-Month Guide
to Planning the Work on
Your Homestead

Cynthia Bombach

Purlin Press
Greensburg, Pennsylvania

Copyright © 2013 by Cynthia Bombach.

All rights reserved. No part of this book may be reproduced or transmitted in any form or by any means, electronic or mechanical, including but not limited to photocopying, recording, Internet transmission or by an information storage and retrieval system, without permission from the publisher, except by a reviewer who may quote brief passages in a review to be included in a magazine, newspaper, website or broadcast.

The information presented in this book has been carefully researched and is accurate to the best of our knowledge. The author and publisher offer no guarantee, nor do they assume any responsibility for any injury, damages or losses incurred during the use of or as a result of following the information presented herein. The author and publisher disclaim any and all liability in connection with the use of this information.

Published in the United States by:
Purlin Press
Contact: homesteadplanner@gmail.com

ISBN 0-9702380-1-0

First edition.

In memory of my dad, Richard Bombach, who instilled in me the heart of a homesteader.

CONTENTS

Introduction... vii

1. **JANUARY**: *Getting Organized*............................... 1
2. **FEBRUARY**: *Preparing for Spring*........................ 13
3. **MARCH**: *Growing Greener*................................... 25
4. **APRIL**: *Cultivating a Healthy Earth*..................... 37
5. **MAY**: *Planting Time*... 49
6. **JUNE**: *Welcoming Summer*..................................59
7. **JULY**: *Celebrating Independence*........................ 69
8. **AUGUST**: *Reaping the Rewards*............................ 79
9. **SEPTEMBER**: *Harvesting Autumn's Bounty*.......... 89
10. **OCTOBER**: *Preparing for Winter*........................ 101
11. **NOVEMBER**: *Giving Thanks*.............................. 111
12. **DECEMBER**: *Resting and Reviewing*.................. 123

Acknowledgements

It would be impossible to adequately thank everyone who has contributed to the information that is gathered in this book. Nevertheless, I owe a debt of gratitude to the homesteaders and farmers who have taught me by word and example, to the authors of the many books, magazine articles and websites about homesteading and farming that I've read, to the Penn State Cooperative Extension and even to the plants and animals from which I have learned so much.

Those who helped me directly with writing and editing this book have my utmost appreciation. My son, Alexander Helzel, gave his keen and honest opinions on layout and editing; Danielle Hager Evans offered her input as to the usefulness of the book; and Joe Zgurzynski of Burgh Bees in Pittsburgh suggested improvements for the bee care sections.

Most of all, I thank God for the experiences that have enabled me to share this information with all of you, the readers for whom I am ever grateful.

INTRODUCTION

When I moved to my first homestead in 1995, I had twenty-five years of experience in gardening and raising animals, but very little practice in planning that work. I had grown up on a small country property on which my family raised vegetables, berries, grapes, chestnuts, rabbits and a pony. There was a lot to do, and for the most part my brother and I did our tasks as they were assigned to us. After college I spent five years living on a small-town lot that had room for nothing more than a few small flower beds, a tiny vegetable patch and two planters on the deck. It didn't take much time or effort to manage such a small amount of work.

Within a few years after moving to our first homestead, however, my husband and I had collected an assortment of chickens, pigs, rabbits, goats and a horse. We had extensive vegetable gardens, a corn field, fruit and nut trees, berry bushes and a pond. It was time for me to get up to speed on managing a homestead, and I had to do it quickly.

I began to read books and magazines about homesteading, and found lots of good advice and how-to instructions. I also learned from my in-laws, who were farmers, and made notes on a calendar as I worked. However, I still struggled to coordinate all my seasonal chores in an organized way. What I really needed was an overall planning guide that would tell me what to do each month.

After spending several years searching for a book that didn't exist, I realized I would have to make one for myself. I labeled twelve folders, one for each month, and began to write the chores for each month on the outsides of the folders. I filled the insides with relevant articles and notes. When I moved from my first homestead to a smaller country property and then to an 11-acre horse farm, I continued to adjust and add to my notes and folders until finally, in the spring of 2011, I began to type them into lists on my computer. As I did so, I realized that other people may benefit from this accumulated knowledge, and this book was born.

The Homestead Planner is meant to be a series of timely reminders to help make your homestead or property more efficient, organized, safe and productive. While the book is intended mainly as a handy planning aid for scheduling the work that needs to be done on a hobby farm or homestead throughout the year, it will also be helpful for women and men who are managing a home for the first time. Anyone who wants to better plan the work around their home or homestead can benefit from at least some of the suggestions. My hope is that the Planner will help others as much as it has helped me.

I've tried to make the Planner as convenient to use as possible. Each month is a self-contained chapter, so that the book can be opened to any month and used immediately without the need to read the previous chapters. Simply start wherever you are and do what seems appropriate for your situation.

Although I have attempted to make the book as comprehensive as possible, The Homestead Planner is by no means a "must-do" list of every single thing to be done on every homestead every year, and it is not a dictator. It is simply a planning guide to help you coordinate your work. You may find that you prefer to do certain tasks at another time of the year, or to skip them entirely. You may do some things one year and not the next. Some tasks won't apply to you at all. For example, some of the items in the "Buildings and Grounds" sections are best suited to those who wish to landscape their property and who enjoy flower gardening; they may not suit some hard-core homesteaders or those with little spare time, just as the livestock suggestions won't apply to most non-homesteading homeowners.

You may have tasks to do that aren't listed in the guide. Because every homestead is different, I've included blank lines for you to fill in with tasks specific to your particular needs. For example, on my first homestead we had to insulate the bathroom stack pipe every December because part of it was exposed on the outside of the house. If we forgot to insulate it, the contents of the pipe would freeze solid, leading to unpleasant backups in the bathroom. We also had to run heat tape to our springhouse because the water line would freeze every January. Hopefully you won't have these problems, but most likely your homestead has a few idiosyncrasies that will require regular attention.

The advice in this book is based on the climate of southwestern Pennsylvania, USDA Hardiness Zone 6. You may have to adjust the timing of certain tasks according to your local climate. Also remember that microclimates can affect your timing as well. If you live in a valley or on a hill, you may get frost earlier or later than average in your area, which would affect the planting and harvest dates of your fruits and vegetables.

Because the Planner is a "when-to" guide rather than a how-to manual, I've kept instructions for each task to a minimum. Consult reputable websites, homesteading publications, your local agricultural extension service or experienced homesteaders in your area when you want more information about how to do any of the tasks mentioned in this guide. Ask your veterinarian if you have doubts or questions about any of the animal care suggestions.

For best results, I use my Homestead Planner in tandem with a wall calendar. I'll record planting and breeding dates, for example, and note memorable homestead activities for each year. I'll also jot down the dates of animal births, peak harvests, and weather phenomena to help me plan for next year. You may also want to transfer some "to-do" items from the Planner to your wall calendar if you know ahead of time when you would like to do them.

Last of all, I've included some suggestions for having fun on the homestead each month. It's a good idea to take some time now and then to enjoy what we've worked so hard to achieve!

****

Your input can improve future editions of The Homestead Planner. Feel free to email your comments and suggestions to homesteadplanner@gmail.com or post them at the Homestead Planner page on Facebook.

1
JANUARY

Getting Organized

Every January our thoughts naturally turn to what we hope to accomplish in the year ahead. For homesteaders this can mean making the garden more productive, trying a new variety of produce or livestock, planning improvements to the property or finding new ways to be more self-sufficient. Whatever you decide to tackle in the coming year, take advantage of this month's long winter evenings to do some research and planning to optimize your chances of success. Be sure to involve anyone who shares the work with you, so you are all on the same page and working toward the same goals. It will be time well spent!

HOUSEHOLD

- If any area of your house feels cluttered, take time to sort through it now, before the spring rush of outdoor work begins.
- Compile tax information. Make an appointment to get taxes done or write a date on the calendar to do them yourself.
- Check credit reports and make any needed corrections. AnnualCreditReport.com is the only source authorized by the federal government to provide free annual reports from each of the three major reporting agencies - Equifax, TransUnion and Experian. You can get a free copy of each report once a year by going to AnnualCreditReport.com, by calling 1-877-322-8228 or by printing the online request form and mailing it to: Annual Credit Report Request Service, P.O. Box 105281, Atlanta, GA 30348-5281. You have the right to request the three reports all at once or at different times during the year, as long as you are ordering a report from each agency only once per year. Staggering the requests every four months allows you to keep a better tab on what's happening with your credit history.
- Purge and organize household files. Bankrate.com, USA.gov and other websites have information about how long to keep important documents and financial records.
- Transfer important dates to new calendar.
- Check heating fuel supplies.
- Check furnace filter; clean or change as needed.
- Test smoke and CO detectors according to manufacturers' instructions.
- Sharpen axes, hatchets, saws and other woodcutting and pruning equipment.
- Disinfect cutting boards. Wipe wooden boards with vinegar, then follow with a fresh paper towel soaked in hydrogen peroxide. To deodorize, rub with the cut side of a lemon. Sanitize plastic boards in dishwasher or wash with mild bleach solution. Maintain a separate cutting board for raw meat.
- If you don't already have one, start a homestead emergency savings account for unexpected veterinary bills, equipment repairs and the like. Boost the account by selling items you no longer need.

JANUARY 3

- Plan and get bids for any building projects or other work to be contracted out this spring.
- _Get A well dug Quote._
- _____
- _____

GARDENING

- Plan this year's garden.
- Make a shopping list of seeds, plants and trees to be ordered from catalogs or bought at a local garden center or nursery. Look for heirloom and open-pollinated varieties, which have seeds that can be harvested every year and used to plant the following year's crop.
- Check seed-starting and gardening supplies; order or purchase needed materials from catalogs or garden centers.
- Plant salad greens indoors under grow lights for late-winter harvest.
- Prune fruit trees.
- _Buy 4 - Bradford Pear trees_
- _____
- _____

HARVESTING/PRESERVING

- Check fruits and vegetables in storage. Promptly use those that are beginning to lose quality, or preserve them by drying or freezing. Do not can produce that is showing signs of damage or spoilage.

- _____

- _____
- _____

FORAGING

- **WARNING: Many wild plants or parts of plants can be toxic! Before consuming any wild foods, be sure you can identify exactly what you are harvesting and that you know how to properly prepare it. Harvest only from clean areas not contaminated with chemical sprays, traffic pollution, animal waste or other toxins. Do not consume large amounts of any wild plant unless you are absolutely sure of its safety.**
- **Before harvesting and consuming wild plant foods, consult an experienced local forager or a reputable foraging field guide such as** *A Field Guide to Edible Wild Plants: Eastern and Central North America* **by Lee Allen Petersen or** *The Complete Guide to Edible Wild Plants* **by the Department of the Army.**

- Pick chickweed for winter salads.
- Dig Jerusalem artichoke tubers. Scrub thoroughly; eat raw or cooked.

- _____
- _____
- _____

LIVESTOCK

- Be sure pastured animals have shelter from wind and precipitation.
- Feed extra hay in very cold or windy weather. The metabolic

processes of digesting roughage help to keep animals warm.
- Check body condition to be sure animals are getting enough feed. For a quick check, feel for ribs beneath hairy winter coats. If ribs are easily felt, your animals need more feed. For a more detailed assessment, refer to one of the many livestock body condition charts available online or through your county Cooperative Extension agent.
- Check at least twice a day to be sure water supplies aren't frozen. Try floats or insulated buckets to help keep water liquid. For a DIY alternative, stack two tires, stuff cavities with straw or hay and nestle a water bucket in the center. Electric buckets and tank heaters should be used only with caution and frequent monitoring. Cords and tank heaters must be inaccessible to animals and should be grounded with a GFCI outlet to prevent electric shock.
- Monitor and treat for lice as needed.
- Check for areas of thick ice in pastures, paddocks and barnyards. Keep animals away from it or scatter sand or sawdust over it for traction.
- Check fences and gates; make any needed repairs.
- Keep animals away from frozen ponds to prevent them from walking onto them and falling in.

- _____
- _____
- _____

Bees

- Build, buy or repair equipment. Clean frames.
- Order bees.
- Check hives on a warm day. Look for activity; if there is none, listen closely in several locations near the thin wood in the handhold of the super. If you still hear nothing, carefully open the hive and see if the bees are alive.
- Feed bees with fondant or sugar brick if necessary.

- _____
- _____

Cattle

- Keep dairy cow udders, teats and bedding dry to prevent frostbite in below-zero temperatures.
- Dry off (stop milking) pregnant dairy cows 60 days before due date. Eliminate grain from cow's diet one week before drying off.
- Trim hooves on dairy cows during drying-off period, about six weeks before calving.

- _____
- _____

Goats

- Assemble kidding supplies including clean towels, iodine and a small cup for navel dipping, blunt scissors and dental floss to cut and tie off umbilical cord if necessary, nasal aspirator syringe, elbow-length plastic gloves, OB lubricant, flashlight, bottles and nipples, feeding tube, frozen colostrum, heat lamp and extra bulb.
- Study goat birthing information if you are inexperienced or need a refresher.
- Give pregnant does CD/T booster vaccines four weeks before kidding due date.
- Decide how you will house and feed goat kids. Prepare housing and feeding equipment. Be prepared to warm newborn kids if necessary with a safely installed heat lamp or other method.
- Check hooves; trim if needed. Avoid stressing heavily pregnant does with hoof trimming.

- _____
- _____

Horses

- Trim hooves.
- Watch for snow and ice buildup in hooves.
- _____
- _____

Pigs

- Give pigs plenty of straw or other bedding in very cold weather. If you use straw or hay for young piglets, chop it first so they can move through it more easily.
- Prepare farrowing stalls for pregnant sows. Have a safely installed heat lamp or other source of warmth available for piglets.
- Prepare farrowing kit, including clean towels, iodine and a small cup for navel dipping, elbow-length plastic gloves, OB lubricant, flashlight, nasal aspirator syringe, colostrum and/or milk replacer or goat milk, feeding tube, human baby bottles and nipples.
- Be very careful of your personal safety around farrowing sows and those raising piglets. Even "pet" sows can become dangerously protective of their young.
- Dip navels of newborn piglets. Warm small, listless piglets under heat lamp if needed. Notch or tattoo ears if needed for identification. Be sure all piglets have a chance to nurse. Give iron shots or supplements to piglets at 3-5 days of age if they do not have access to soil in their pen. Clip tips of sharp canine teeth ("needle teeth") if desired. Castrate males at about one week of age.
- _____
- _____

Poultry

- Collect eggs several times a day if temperatures are below freezing.
- Clean nest boxes and line with fresh hay or shavings.
- Check twice a day to be sure water supplies aren't frozen.
- Keep bedding clean and dry.
- Check coop for drafts and correct if necessary, but do not make the building airtight.
- Cracked corn scattered on floor will keep hens busy and provide extra calories.
- Give birds fresh greens, vegetable and fruit trimmings for a winter treat, but avoid giving them raw potato peelings or strong-tasting foods such as onions.

- _____
- _____

Rabbits

- Be sure rabbit housing has shelter from drafts and precipitation. Add bedding as needed to keep rabbits warm.
- Check water twice a day.

- _____
- _____

Sheep

- Prepare for spring lambing. Gently shear pregnant ewes 30-45 days before due date if weather will be mild. Withhold food and water for 12 hours prior to shearing. Provide adequate shelter and extra feed for shorn ewes in cold weather. As an alternative to shearing, "crotch" ewes by clipping only the udder and groin areas prior to lambing.

JANUARY 9

- Prepare lambing pens, if using. Be prepared to warm newborn lambs if necessary with a safely installed heat lamp or other method.
- Gather supplies for lambing kit, including elbow-length plastic gloves, OB lubricant, towels, nasal aspirator syringe, blunt scissors and dental floss to cut and tie off umbilical cord if necessary, iodine and small cup to dip navel, bottles and nipples, frozen colostrum, milk replacer, feeding tube and head lamp or flashlight.
- Review lambing information in books or online if you are inexperienced or need a refresher.
- Give pregnant ewes CD/T booster shot about four weeks before due date.
- Dock lambs' tails at 2-3 days of age. Castrate males within the first week of life or when both testicles have descended into scrotum. Castration is not necessary for lambs to be slaughtered before six months of age, as long as they are separated from ewes before they are four months old.
- Check hooves; trim if needed. Avoid stressing heavily pregnant ewes with hoof trimming.
- _____
- _____

PETS

- Brush dogs frequently to reduce hair buildup in the house.
- Wash pet bedding and dishes.
- Trim nails.

- _____
- _____
- _____

BUILDINGS & GROUNDS

- Prune summer and fall-blooming trees, shrubs and vines. Cut up small branches for kindling.
- Cut down tree branches damaged by winter storms.
- Continue cutting trees and brush as needed. Consult county Extension agent for woodlot management plan.
- Check trees and shrubs for rodent and rabbit damage. Install tree guards around trunks if needed. Keep mulch several inches away from bases of trunks to reduce rodent hiding places.
- Water young evergreens during January thaw.
- Brush heavy snow buildup off evergreens and other trees and shrubs. Don't try to break off ice – doing so could damage branches.

- _____
- _____
- _____

MACHINERY & EQUIPMENT

- Check fluid levels and tire pressure as instructed in owners' manuals.
- Wash vehicles regularly to rid them of road salt.
- Schedule the year's routine maintenance tasks, such as oil and filter changes and tire rotations, according to owners' manuals.

- _____
- _____
- _____

ENJOYMENT

- The key to tolerating a long, cold winter is to get out there and enjoy it! Try a new winter activity such as snowshoeing, bird watching or outdoor photography.
- Brighten up a winter evening by hosting a night of old-fashioned parlor games like charades and hot potato for family or friends. Find games by searching for "parlor games" online.

- _____
- _____
- _____

2
FEBRUARY

Preparing for Spring

Just when it seems that winter will never end, February arrives and brings with it the first hints of spring on the homestead. Newborn sheep and goats appear beside their mothers, early bulbs may peek up through the snow, and maple sap begins to flow. Late in the month, we can begin starting seeds indoors for spring planting in the garden. The feel of soil (or soilless mix) in our hands and the sight of green shoots sprouting will give us hope that spring is just around the corner!

HOUSEHOLD

- Update insurance policies. Ask your agent for information about liability insurance relating to livestock, ponds and hazardous structures on your property.
- Assess and update home security measures including outdoor lighting, door and window locks, visibility, alarms, fences, home-monitoring systems and/or watchdogs.
- Enroll the family in a self-defense class. Contact your local police department or martial arts studio to find a class.
- If you keep firearms in the house, be sure that every adult family member knows how to handle and use them safely, and that they are safeguarded from children.
- Make sure that everyone in the house knows how to turn off the main water, gas and electric supplies.
- Have a fire drill. Try to plan at least two escape routes from each room. Decide on an outdoor meeting place in the event of a fire. If you live in a two-story or taller house, purchase a collapsible fire escape ladder that hooks over a windowsill. Keep it in an easily accessible upstairs location.
- Place fire extinguishers in the kitchen, near the bedrooms, at the top of the basement stairwell, in the basement, in the garage and in the workshop. Check older extinguishers for viability according to manufacturers' directions. Promptly replace expired extinguishers or have them professionally recharged.
- Take a basic first aid/CPR class. Contact your local Red Cross or community college for class availability.
- Post written directions to your homestead in the kitchen or near a landline phone for use when calling for help in an emergency.
- Clean lint from clothes dryer vents, including exterior vent.
- Vacuum vents and coils on refrigerators and freezers. Unplug appliances before working on them.
- Check heating fuel supplies.
- Check furnace filter; clean or change as needed.
- Test smoke and CO detectors according to manufacturers' instructions.
- Do spring cleaning now to get it out of the way before heavy outdoor work begins.

FEBRUARY   **15**

- _____
- _____
- _____

GARDENING

- Plan garden layout if not done already. Study crop rotation to get maximum production from garden beds.
- Schedule seed-starting and planting dates according to seed packet directions and local frost dates.
- Organize seed packets. Check old seeds for viability by placing between layers of damp paper towels. Keep moist during the germination period stated on seed packet. If seeds don't sprout, discard them and use fresh replacements.
- Purchase and prepare seed-starting items, including sterilized soil or soilless mix, peat pots, cell packs or homemade containers, heat mats (purchased or homemade; instructions for various DIY heat mats can be found by searching online) and grow lights.
- Purchase any needed garden supplies and tools.
- Take advantage of down time to build new compost bins, cold frames, raised beds and other garden structures.
- Toward the end of the month, start seeds of celery, leeks and onions indoors under grow lights.
- Pinch and prune potted herbs and greenhouse plants.
- Prune grape vines.
- Finish pruning fruit trees.
- _Sun Chokes-_____
- _____
- _____

HARVESTING/PRESERVING

- _____
- _____
- _____

FORAGING

- **WARNING: Many wild plants or parts of plants can be toxic! Before consuming any wild foods, be sure you can identify exactly what you are harvesting and that you know how to properly prepare it. Harvest only from clean sites not contaminated with chemical sprays, traffic pollution, animal waste or other toxins. Do not consume large amounts of any wild plant unless you are absolutely sure of its safety.**
- **Before harvesting and consuming wild plant foods, consult an experienced local forager or a reputable foraging field guide such as one of the following:** *A Field Guide to Edible Wild Plants: Eastern and Central North America* **by Lee Allen Petersen or** *The Complete Guide to Edible Wild Plants* **by the Department of the Army.**

- Tap maple trees when daytime temperatures are above 32 degrees and nighttime temperatures are below 32.
- Dig Jerusalem artichoke tubers. Scrub thoroughly; eat raw or cooked.

- _____
- _____
- _____

LIVESTOCK

- Feed extra hay in very cold or windy weather. The metabolic processes of digesting roughage help to keep animals warm.
- Check body condition to be sure animals are getting enough feed. Feel for ribs beneath hairy winter coats. If ribs are easily felt, your animals need more feed. For a more detailed assessment, refer to one of the many livestock body condition charts available online or through your county Cooperative Extension agent.
- Check at least twice a day to be sure water supplies aren't frozen. Try floats or insulated buckets to help keep water liquid. For a DIY alternative, stack two tires, stuff cavities with straw or hay and nestle a water bucket in the center. Electric buckets and tank heaters should be used only with caution and frequent monitoring. Cords and tank heaters must be inaccessible to animals and should be grounded with a GFCI outlet to prevent electric shock.
- Check fences and gates; repair as needed.
- Check hay and bedding supplies; order more if necessary.
- Don't let animals stand around in mud. Contact your county Extension agent for help correcting chronically muddy areas, and/or move the animals to a different area.
- Schedule spring vaccinations. Ask your veterinarian about vaccinations for pregnant and newborn animals.

- _____
- _____
- _____

Bees

- Feed bees with fondant or sugar brick if necessary.
- Begin to remove entrance reducers as bee activity increases.

- _____
- _____

Cattle

- Keep udders, teats and bedding dry to prevent frostbite in below-zero temperatures.
- If you have cows ready to calve, check calving supplies. Include a halter and lead rope for the cow, obstetrical gloves and lubricant, calving chains, towels, dull scissors and dental floss for cutting and tying off umbilical cord if necessary, iodine spray or iodine and dipping cup for navel, nasal aspirator syringe, veterinary thermometer, flashlight, ear tags if using, bottle and lamb nipple. Have a clean, dry stall available in case it's needed to warm calf.
- Study calving information online or in cattle care books if you are inexperienced or need a refresher.
- Be careful of your personal safety when approaching or handling cows that are calving or that have calves. Cows that need assistance with calving should be placed in a calving chute for the handler's safety. A properly constructed chute includes a head gate that restrains the cow while allowing her enough freedom to stand or lie down comfortably. The chute should have adjustable side panels that allow the handler or veterinarian safe access to the cow as well as allowing the calf to nurse. Plans for simple homemade chutes are available from many county Extension offices.
- Castrate bull calves when both testicles have descended into scrotum. Dehorn calves if desired.
- Dry off (stop milking) pregnant dairy cows 60 days before due date. Eliminate grain from her diet one week before drying off.
- _____
- _____

Goats

- September-bred does will be kidding this month. If not done previously, assemble kidding supplies including clean towels, iodine and a small cup for navel dipping, blunt scissors and dental

floss to cut and tie off umbilical cord if necessary, elbow-length plastic gloves, OB lubricant, flashlight, nasal aspirator syringe, bottles and nipples, feeding tube, frozen colostrum, heat lamp and extra bulb.
- Study goat birthing information if you are inexperienced or need a refresher.
- Give pregnant does CD/T booster vaccines four weeks before kidding due date.
- Decide how you will house and feed goat kids. Prepare housing and feeding equipment. Be prepared to warm newborn kids if necessary with a safely installed heat lamp or other method.
- Disbud kids before two weeks of age, castrate before four weeks.
- Handle kids frequently to keep them manageable.
- _____
- _____

Horses

- Watch for snow and ice buildup in hooves.
- _____
- _____

Pigs

- Begin now to locate a source for feeder pigs, especially if supply is limited in your area.
- Prepare pens, feeders and waterers if you plan to purchase feeder pigs.
- Prepare farrowing stalls for pregnant sows. Have a safely installed heat lamp or other source of warmth available for piglets.
- Assemble farrowing kit, including clean towels, iodine and a small cup for navel dipping, elbow-length plastic gloves, OB lubricant, flashlight, nasal aspirator syringe, colostrum and/or

milk replacer or goat milk, feeding tube, human baby bottles and nipples.
- Be very careful of your personal safety around farrowing sows and those raising piglets. Even "pet" sows can become dangerously protective of their young.
- Dip navels of newborn piglets. Warm small, listless piglets under heat lamp if needed. Notch or tattoo ears if needed for identification. Be sure all piglets have a chance to nurse. Give iron shots or supplements to piglets at 3-5 days of age if they do not have access to soil in their pen. Clip tips of sharp canine teeth ("needle teeth") if desired. Castrate males at about one week of age.
- Butcher fall-born hogs for Easter hams.
- _____
- _____

Poultry

- Order chicks, ducklings, guinea keets and goslings from catalog or farm store for shipment in March, April or May. Order turkey poults to arrive in late May.
- Prepare brooders for hatchlings. Find instructions for homemade designs in poultry books or online, or purchase a brooder from a poultry supply catalog or farm store.
- Be sure to get the appropriate poultry starter feed for each type of bird.
- Purchase heat lamps, feeders and waterers as needed.
- _____
- _____

Rabbits

- Be sure rabbits have dry, draft-free housing.
- Check water at least twice a day when temperatures are below freezing.
- Purchase new breeding stock if needed.
- Breed does that are at least nine months old. Breeding two or more does at once will allow for fostering of excess or rejected young.
- On the 28th day after breeding, place a nest box in each pregnant doe's cage.

- _____
- _____

Sheep

- Prepare for spring lambing if not done in January. Gently shear pregnant ewes 30-45 days before due date if weather will be mild. Withhold food and water for 12 hours prior to shearing. Provide adequate shelter and extra feed for shorn ewes in cold weather. As an alternative to shearing, "crotch" ewes by clipping only the udder and groin areas prior to lambing.
- Prepare lambing pens, if using. Be prepared to warm newborn lambs if necessary with a safely installed heat lamp or other method.
- Gather supplies for lambing kit if not done previously, including elbow-length plastic gloves, OB lubricant, towels, nasal aspirator syringe, blunt scissors and dental floss to cut and tie off umbilical cord if necessary, iodine and small cup to dip navel, bottles and nipples, frozen colostrum, milk replacer, feeding tube and head lamp or flashlight.
- Give pregnant ewes CD/T booster shot about four weeks before due date.
- Dock lambs' tails at 2-3 days of age. Castrate males within the first week of life or when both testicles have descended into

scrotum. Castration is not necessary for lambs to be slaughtered before six months of age, as long as they are separated from ewes before they are four months old.

- _____
- _____

PETS

- Trim nails.
- _____
- _____
- _____

BUILDINGS & GROUNDS

- Finish major cutting of brush and firewood.
- Check trees and shrubs for winter damage; prune off broken branches.
- Pinch and prune greenhouse plants as necessary.
- _____
- _____
- _____

MACHINERY & EQUIPMENT

- Check fluids and tire pressure according to owners' manuals.
- Wash vehicles and clean floor mats to keep road salt from corroding surfaces.

- _____
- _____
- _____

ENJOYMENT

- Try out some new recipes for homemade pizza or baked goods to warm the kitchen on a cold evening.
- Enjoy a snowy day by building a snowman with the kids.
- Learn to identify the many animal tracks you find in the snow. You might be surprised to find out which animals have been visiting your homestead at night!

- _____
- _____
- _____

3
MARCH

Growing Greener

This month the homestead begins to turn green again… except where it's brown. The weather that brings us the wonders of springtime also brings the not-so-wonderful mud. Whether caked on the feet of livestock or carried into the house on barn boots and dog paws, mud seems to be everywhere. A pasture management plan and a boot scraper by the back door will help until things dry out.

HOUSEHOLD

- Check heating fuel supply. Make sure firewood is staying dry.
- Check furnace filter; clean or change as needed.
- Remove excess ashes from wood burners and fireplaces.
- Put new batteries in smoke and carbon monoxide detectors. Replace outdated detectors.
- Clean and organize mud room or family entrance area.
- Groom and pinch back houseplants. Apply organic fertilizer to plants showing signs of new growth.
- To reduce floor-cleaning chores, place a heavy-duty doormat outside each entrance door and a washable, nonslip cloth mat inside each door. Clean mats frequently.
- Plan what your family will do in case of a tornado. Designate a lower-level interior room as a sheltering place. Have shoes and any kind of helmets handy, along with an old mattress, sturdy workbench or other protection to hide under. Decide what you will do with pets if a tornado threatens.

- _____
- _____
- _____

GARDENING

- Finalize garden plans.
- Test soil and amend if necessary. Testing kits and information are available through garden centers, seed catalogs and county Extension agents.
- Buy any needed garden supplies. Consider using floating row cover over garden beds to repel pests and protect plants from light frosts.
- Start pepper seeds at the beginning of the month.
- Early in the month, insert toothpicks into the sides of one or two fat, healthy sweet potatoes and suspend each one in a jar of water for sprouting. Alternatively, omit toothpicks and simply bury

sweet potatoes halfway in a pan of soil. Keep indoors in a warm, sunny place. When leafy sprouts are 4-5" tall, gently break them off and place in a small jar of water until they root. Plant well-rooted slips in peat pots or small cups to grow until late-May transplanting in garden.
- Finish fruit tree pruning and dormant oil spraying by early March.
- Start seeds of broccoli, cabbage, cauliflower and Brussels sprouts in mid-March.
- Plant lettuce, spinach, Swiss chard and radishes midmonth, or when soil is workable.
- Start tomato seeds in mid-to-late March, aiming to transplant seedlings twice into larger pots before planting outside in mid-to-late May. Plant seedlings slightly deeper each time to strengthen the stem and root systems.
- Prune raspberry canes by removing suckers and spindly stems, leaving four to six canes per hill. Then trim off top one-fourth of remaining canes. Or, to extend the season, leave half the canes alone and cut the top half off the others. Dispose of all cuttings in the trash to reduce the spread of pests and disease organisms.
- Put aged manure on garden beds, rhubarb plants, asparagus, strawberries, raspberries, garlic and young grape vines.
- Build raised beds and fill with soil or compost mixture.
- Turn under winter rye cover on existing raised beds 3-4 weeks before planting.
- Cover beds designated for onion seedlings and sets with clear plastic to warm them before the planned mid-April planting date.
- Put up fences around gardens to keep out rabbits, deer, groundhogs and other pests. Hang sweaty work shirts or sprinkle human or dog urine around fence to further deter pests.

- _____
- _____
- _____

28   THE COMPLETE HOMESTEAD PLANNER

HARVESTING/PRESERVING

- Make maple syrup.
- Harvest horseradish and parsnips.

- _____
- _____
- _____

FORAGING

• **WARNING: Many wild plants or parts of plants can be toxic! Before consuming any wild foods, be sure you can identify exactly what you are harvesting and that you know how to properly prepare it. Harvest only from clean sites not contaminated with chemical sprays, traffic pollution, animal waste or other toxins. Do not consume large amounts of any wild plant unless you are absolutely sure of its safety.**
• **Before harvesting and consuming wild plant foods, consult an experienced local forager or a reputable foraging field guide such as one of the following:** *A Field Guide to Edible Wild Plants: Eastern and Central North America* **by Lee Allen Petersen or** *The Complete Guide to Edible Wild Plants* **by the Department of the Army.**

- Pick dandelion leaves, chickweed and corn salad (mache) for spring salads.
- Look for watercress in clean springs and streams well away from livestock areas. Rinse, soak for 30 minutes in two quarts of water with 2 tablespoons vinegar added to kill water-borne parasites, then rinse well before using. Eat fresh in salads and sandwiches.
- Dig Jerusalem artichoke tubers. Scrub well; eat raw or cooked.

LIVESTOCK

- Keep animals out of muddy areas when possible. Consult your local conservation district or Cooperative Extension office for advice and help in managing mud-prone areas.
- Lay out rotational grazing paddocks. Add new fencing, water sources and shelters as needed.
- Scatter grass and/or clover seed on thinning pastures early in the month. Freeze-thaw cycles will work seeds into the soil. Keep animals off pasture for 3-4 months after seeding. Get advice from county Extension agent for best practices in your area.
- Keep animals off pasture until grass is 6-8 inches tall.
- Check pasture fences and gates; make needed repairs.
- Make sure cattle, goats and sheep have access to a high-quality mineral block or supplement with magnesium beginning 4-6 weeks before pasture turnout to help avoid grass tetany, a potentially deadly metabolic disorder caused by magnesium deficiency.
- Develop and implement manure management plan. Keep manure pile 100-200' from wells, springs, streams and ponds. Avoid spreading manure on frozen ground.
- Handle newborn and young livestock frequently to help make them more manageable. Beware of sows, cows and other protective livestock when handling their offspring.

Bees

- Feed bees with 1:1 sugar syrup if their honey supply is low.
- Remove entrance reducers from bee boxes.
- Get supers, empty frames and new foundation ready for spring use.
- Order any needed equipment.
- Provide a convenient source of fresh water if none is available.
- Scout for new apiary sites. Look for sites that are easily accessible, sheltered from the wind, that face east or south and have plentiful forage nearby.

- _____
- _____

Cattle

- If you have cows ready to calve, check calving supplies. Include a halter and lead rope for the cow, obstetrical gloves and lubricant, calving chains, towels, dull scissors and dental floss for cutting and tying off umbilical cord if necessary, iodine spray or iodine and dipping cup for navel, nasal aspirator syringe, veterinary thermometer, flashlight, ear tags if using, bottle and lamb nipple. Have a clean, dry stall available in case it's needed to warm calf.
- Study calving information if you are inexperienced or need a refresher.
- Castrate bull calves when both testicles have descended into scrotum. Dehorn calves if desired.
- Be careful of your personal safety when approaching or handling cows that are calving or that have calves. Cows that need assistance with calving should be placed in a calving chute for the handler's safety. A properly constructed chute includes a head gate that restrains the cow while allowing her enough freedom to stand or lie down comfortably. The chute should have adjustable side panels that allow the handler or veterinarian safe access to the cow as well as allowing the calf to nurse. Plans for

simple homemade chutes are available from many county Extension offices. Contact your county Extension agent for more information.
- Dry off (stop milking) pregnant dairy cows 60 days before due date. Eliminate grain from their diets one week before drying-off date.

- _____
- _____

Goats

- October-bred does will begin kidding. If not done previously, assemble kidding supplies including clean towels, iodine and a small cup for navel dipping, blunt scissors and dental floss to cut and tie off umbilical cord if necessary, elbow-length plastic gloves, OB lubricant, flashlight with fresh batteries, nasal aspirator syringe, bottles and nipples, frozen colostrum, a feeding tube and syringe, heat lamp and extra bulb.
- Study goat birthing information if you are inexperienced or need a refresher.
- Give pregnant does CD/T booster vaccines four weeks before kidding due date.
- Decide how you will house and feed goat kids. Prepare housing and feeding equipment. Be prepared to warm newborn kids if necessary with a safely installed heat lamp or other method.
- Disbud kids before two weeks of age, castrate before four weeks.
- Handle kids frequently to keep them manageable.
- Market kids that you don't plan to keep.
- Check hooves; trim if needed. Avoid stressing heavily pregnant does with hoof trimming.

- _____
- _____

Horses

- Trim hooves.
- Resume riding or other work gradually if horse has been idle over winter.
- _____
- _____

Pigs

- Prepare pen and/or pasture for feeder pigs if not done yet.
- Prepare farrowing stalls for pregnant sows. Have a safely installed heat lamp or other source of warmth available for piglets.
- Assemble farrowing kit, including clean towels, iodine and a small cup for navel dipping, elbow-length plastic gloves, OB lubricant, flashlight, nasal aspirator syringe, colostrum and/or milk replacer or goat milk, feeding tube, human baby bottles and nipples.
- Be very careful of your personal safety around farrowing sows and those raising piglets. Even "pet" sows can become dangerously protective of their young.
- Dip navels of newborn piglets. Warm small, listless piglets under heat lamp if needed. Notch or tattoo ears if desired for identification. Be sure all piglets have a chance to nurse. Give iron shots or supplements to piglets at 3-5 days of age if they do not have access to soil in their pen. Clip tips of sharp canine teeth ("needle teeth") if desired. Castrate males at about one week of age.
- Sell home-raised feeder pigs now, when prices are traditionally highest.
- _____
- _____

MARCH 33

Poultry

- Roosters often become more aggressive in spring. Monitor hens for feather loss and damage due to aggressive mating behavior. Separate rooster or protect hens with purchased or homemade "saddles." Be ready to defend yourself when an aggressive rooster is nearby!
- Do not allow very young waterfowl to swim on ponds or lakes where predators such as bass and snapping turtles are present.

- _____
- _____

Rabbits

- Examine litters of newborn bunnies; remove dead ones and place excess or rejected ones with foster doe if available.

- _____
- _____

Sheep

- Prepare for spring lambing if not done yet. Gently shear pregnant ewes 30-45 days before due date if weather will be mild. Withhold food and water for 12 hours prior to shearing. Provide adequate shelter and extra feed for shorn ewes in cold weather. As an alternative to shearing, "crotch" ewes by clipping only the udder and groin areas prior to lambing.
- Prepare lambing pens, if using. Be prepared to warm newborn lambs if necessary with a safely installed heat lamp or other method.
- Gather supplies for lambing kit if not done previously, including elbow-length plastic gloves, OB lubricant, towels, nasal aspirator syringe, blunt scissors and dental floss to cut and tie off umbilical cord if necessary, iodine and small cup to dip navel, bottles and

nipples, frozen colostrum, milk replacer, feeding tube and head lamp or flashlight.
- Give pregnant ewes CD/T booster shot about four weeks before due date.
- Dock lambs' tails at 2-3 days of age. Castrate males within the first week of life or when both testicles have descended into scrotum. Castration is not necessary for lambs to be slaughtered before six months of age, as long as they are separated from ewes before they are four months old.
- Protect pasture-born lambs from predators with fencing and/or guardian animals.
- Check hooves; trim if needed. Avoid stressing heavily pregnant ewes with hoof trimming.
- Shear sheep as soon as weather permits, preferably before fly season.

- _____
- _____

PETS

- Trim nails.
- Groom shedding dogs frequently to minimize hair in house.
- Wash pet beds and dishes.

- _____
- _____
- _____

BUILDINGS & GROUNDS

- Check for drainage issues with foundations, driveways and other areas. Plan to make needed corrections to grading when the ground is dry.
- Check roof gutters for ice damage. Clean gutters and downspouts.

- Finish trimming trees and shrubs damaged by heavy snow or ice.
- Cut back dead perennials and ornamental grasses left standing over winter.
- Late in the month, cut back butterfly bushes to about 18."

- _____
- _____
- _____

Ponds

- Remove winter debris.
- Refill small artificial ponds. Provide stack of rocks or floating objects for frogs to sit on.

- _____
- _____

MACHINERY & EQUIPMENT

- Check fluid levels and tire pressure according to owners' manuals.
- Wash vehicles.
- Get lawn mower and lawn tractor ready for spring work: charge battery; refill gas tank; clean, sharpen and balance blades if not done in the fall.
- Sell unneeded equipment.

- _____
- _____
- _____

ENJOYMENT

- Fly a kite!
- Put up bird houses and nest boxes.
- Keep feeding the birds – natural food supplies are now at their lowest level of the year.
- Take a few minutes in the morning to listen to the bird chorus, whether wild or domestic. Nothing greets the morning like chirping songbirds or the sound of a hen announcing the arrival of a freshly laid egg!

- _____
- _____
- _____

4

APRIL

Cultivating a Healthy Earth

This month's observance of Earth Day is a reminder to be careful of what we put into our environment, because eventually it all comes back to us through the air we breathe, the water we drink and the plants and animals that we eat. It's not always possible to avoid chemicals completely, but we can use mostly natural and organic methods of cleaning, gardening, animal care and pest control. Baking soda and vinegar are effective household cleaners. Natural soil amendments such as manure and compost will improve the fertility and quality of our gardens. Raising animals in a healthy, natural environment will reduce the need for medications. Floating row covers, hand-picking and blasts of water from a garden hose will help keep pests off our garden produce. Caring for the homestead with natural methods will help ensure that our land will continue to provide a healthy life for us and for future generations.

38 THE COMPLETE HOMESTEAD PLANNER

HOUSEHOLD

- Check furnace filter; clean or change as needed.
- Test smoke and CO detectors according to manufacturers' instructions.
- Clean and sanitize springs and wells as recommended by your county Extension agent or department of environmental resources. Test water for bacteria and contaminants two weeks after cleaning.
- Remove heat tape and extra insulation from springhouses and wellheads.
- Install low-flow faucet and shower heads with a shut-off lever, which allow you to stop and restart the flow of water at the same temperature without readjusting the knobs.
- Drain and refill water heater. Consult owner's manual or search online for detailed instructions.
- Fix leaky faucets, including outdoor faucets.
- Clean sediment from washing machine water lines. Read instruction manual for your machine. General procedure is to unplug the machine from the wall outlet, then shut off water valves and disconnect supply hoses to machine. Have a bucket handy to drain excess water into. Suck out the machine's hoses with a wet/dry vacuum, tapping or jiggling them as you do so to loosen sediment. Reconnect hoses, turn on water valves and plug in machine.
- Clean sump pump and pit. Consult owner's manual or search online for detailed instructions. General procedure is to turn off electricity to pump and remove it from pit. Place pump in a bucket and take it outside to clean it. Siphon water from pit and scrub off sludge. Return cleaned pump to pit and reconnect it, then restore power.
- Take down and clean heavy window coverings and storm windows. Wash and store for the summer. Clean windows with a mixture of equal parts vinegar and water. Install window screens and enjoy the fresh spring air.
- Vacuum or wipe down walls and ceilings to remove dust, cobwebs and fireplace ash.
- Check for insect or wasp infestations in attic; treat as necessary.

- Look for signs of termites and carpenter ants around foundation and in basements and crawl spaces. Besides chewing tunnels in wood, termites build mud tunnels between the ground and the wooden parts of the house. Carpenter ants leave piles of fine sawdust under areas of chewed wood. Search online for natural pest control hints and/or hire an exterminator that specializes in natural and low-toxicity methods.
- Clean and check gutters and downspouts. Be sure runoff drains away from foundation. Check foundation for cracks.
- Go up in the attic after a heavy or prolonged rainfall and inspect for roof leaks.
- Prepare a flood evacuation plan for your family. Decide at what point you will leave and where you will go if a flood threatens your house. Decide what you will do with pets and livestock in case of a flood.
- Uncover or put out outdoor furniture, grills, toys and birdbaths.
- Prepare grill or outdoor fireplace for use.
- Open outdoor faucets, attach hoses and check irrigation systems.
- Check septic tank and have it pumped out if solids make up one-third to one-half the contents.

- _____
- _____
- _____

GARDENING

- Start tomato seeds early in the month if not done in late March. Plan to transplant seedlings twice into larger pots before planting outside in late May. Plant seedlings slightly deeper each time to strengthen the stem and root systems.
- Continue to carefully break off sweet potato sprouts from the potato when they reach 4-5" tall. Place each sprout in a small cup or jar of water until well rooted, then plant in cups or pan of soil amended with a little bone meal and compost. Water regularly, but keep soil on the dry side.

40 THE COMPLETE HOMESTEAD PLANNER

- Prepare all garden beds for planting. Turn under winter rye cover or remove mulch. Loosen soil and pull out any weeds. Add any needed amendments as indicated by soil test.
- Plant carrots, lettuce, Swiss chard, spinach, onions, parsnips, turnips, peas and radishes.
- Mid-month or later, plant broccoli, cauliflower and cabbage seedlings in garden. Protect broccoli and cauliflower from cold temperatures (under 50 degrees) by covering each one with a milk jug that has had its cap and bottom removed.
- Plant potatoes between mid-April and June. Buy seed potatoes or cut a healthy potato in large pieces with an eye in each piece. Let sit to dry slightly for 24 hours, then plant 6-8" deep in garden or bury under straw or compost. Mound up dirt or mulch as plant grows.
- Plant asparagus and strawberry plants.
- Plant blueberry bushes.
- Weed raspberry patch and mulch with compost or leaves.
- _____
- _____
- _____

HARVESTING/PRESERVING

- Harvest rhubarb; freeze, can or make sauce, jam, pie or cake.
- Pick radishes, spinach and early salad greens.
- Harvest parsnips left in ground over winter.
- _____
- _____
- _____

FORAGING

- **WARNING:** Many wild plants or parts of plants can be toxic! Before consuming any wild foods, be sure you can identify exactly what you are harvesting and that you know how to properly prepare it. Harvest only from clean sites not contaminated with chemical sprays, traffic pollution, animal waste or other toxins. Do not consume large amounts of any wild plant unless you are absolutely sure of its safety.
- Before harvesting and consuming wild plant foods, consult an experienced local forager or a reputable foraging field guide such as one of the following: *A Field Guide to Edible Wild Plants: Eastern and Central North America* **by Lee Allen Petersen or** *The Complete Guide to Edible Wild Plants* **by the Department of the Army.**

- Look for morel mushrooms when apple trees and violets are blooming, around the end of the month. Be sure you can identify morels with 100% certainty before picking! Cook and use like cultivated mushrooms.
- Pick dandelion greens and flower buds. Use fresh in salads with hot bacon or mayonnaise dressing, or cook like spinach.
- Pick chickweed and corn salad (mache) for spring salads.
- Look for watercress in clean springs and streams well away from livestock areas. Rinse, soak for 30 minutes in two quarts of water with 2 tablespoons vinegar added to kill water-borne parasites, then rinse well before using. Eat fresh in salads and sandwiches.
- Pick plantain leaves. Eat fresh or cooked.
- Harvest leaves, young stems and roots of redroot pigweed. Use leaves and stems as fresh or cooked vegetable; cook and mash roots as you would potatoes.
- Harvest young dock leaves. Cook like spinach.
- Harvest young chicory plants to use as cooked vegetable.
- Bulrush shoots can be eaten raw or cooked.
- Daylily buds and shoots can be eaten raw or lightly cooked.
- Harvest young mint leaves; use fresh or dried.

- Dig Jerusalem artichoke tubers. Scrub thoroughly; eat raw or cooked.

- _____
- _____
- _____

LIVESTOCK

- Near the end of the month, give all livestock housing a thorough spring cleaning. Sweep down cobwebs, remove old bedding down to bare floors and open windows and doors to let buildings air out on a sunny day. Scrub and disinfect feeders and waterers. Check doors, gates and windows for safety and security. Make any other needed repairs.
- De-clutter and organize tool and equipment storage areas. Sell, trade or give away unneeded items. Consider using any profits to beef up the homestead emergency fund.
- Clean and organize feed storage areas. Storing grain and feed in an old chest freezer helps keep it dry and protects it from rodents, insects and livestock.
- Check all fences and gates in pastures and paddocks; repair as needed.
- Look over pastures for hazards including groundhog holes, mud pits, trash thrown into field, fallen trees and poisonous plants. Immediately remove any fallen cherry tree branches, as the wilted leaves are poisonous to livestock.
- Introduce animals to pasture gradually. Feed hay first and let them graze half an hour at a time, increasing the length day by day until they are grazing as long as you want them to.
- Watch for bloat when cattle, goats and sheep begin grazing lush pastures. A bloated animal will have a swollen left side and will become increasingly distressed as its breathing is hindered by the expansion of gas in the rumen. Bloat is a medical emergency and must be treated immediately to save the animal's life. Study bloat treatment and prevention information before it happens by

searching online or consulting livestock books, Cooperative Extension services or your veterinarian.
- Study and practice rotational grazing systems.
- Monitor and treat for external parasites as needed.

- _____
- _____
- _____

Bees

- Get honey equipment and supplies ready.
- Remove all entrance reducers and winter mouse guards.
- Install supers as honey flow begins.
- Continue to feed 1:1 sugar syrup if necessary until natural nectar sources are consistent and plentiful.

- _____
- _____

Cattle

- Buy weaned calves if desired for beef or dairy production.
- If you have cows ready to calve, check calving supplies. Include a halter and lead rope for the cow, obstetrical gloves and lubricant, calving chains, towels, dull scissors and dental floss for cutting and tying off umbilical cord if necessary, iodine spray or iodine and dipping cup for navel, nasal aspirator syringe, veterinary thermometer, flashlight, ear tags if using, bottle and lamb nipple. Have a clean, dry stall available in case it's needed to warm calf.
- Study calving information if you are inexperienced or need a refresher.
- Castrate bull calves when both testicles have descended into scrotum. Dehorn calves if desired.

- Be careful of your personal safety when approaching or handling cows that are calving or that have calves. Cows that need assistance with calving should be placed in a calving chute for the handler's safety. A properly constructed chute includes a head gate that restrains the cow while allowing her enough freedom to stand or lie down comfortably. The chute should have adjustable side panels that allow the handler or veterinarian safe access to the cow as well as allowing the calf to nurse. Plans for simple homemade chutes are available from many county Extension offices. Contact your county Extension agent for more information.

- _____
- _____

Goats

- _____
- _____

Horses

- Adjust feed rations as grazing begins and horses become more active.
- Clean and organize tack room.

- _____
- _____

Pigs

- Purchase feeder pigs. Look for healthy pigs that are 6-8 weeks old and weigh 40-60 pounds. Pigs should already have been wormed, vaccinated and castrated. Avoid unhealthy pigs, runts and

uncastrated males. FYI: Gilts (female pigs) produce leaner meat than barrows (castrated males).
- Sell home-raised feeder pigs.

- _____
- _____

Poultry

- Give poultry housing a thorough spring cleaning. With the birds outside or confined elsewhere on a sunny day, sweep down cobwebs, remove all litter and nest material, scrub feeders and waterers, and let everything air out for the day. Check building for needed repairs, watching for rodent holes and songbird entry points. At the end of the day, put in fresh litter and nest material, refill feeders and waterers and return birds to the building.
- Prepare chicken tractors or range shelters for pastured poultry. Be sure to protect birds from predators with wire enclosures, electrified netting or other means.
- Begin pasturing poultry if weather is mild and grass is plentiful.
- Do not allow very young waterfowl to swim on ponds or lakes where predators such as bass and snapping turtles are present.

- _____
- _____

Rabbits

- Remove young from doe's hutch at six weeks of age.
- Clean cages, hutches and nest boxes. Scrub feeders and waterers.

- _____
- _____

Sheep

- Dock lambs' tails at 2-3 days of age. Castrate males within the first week of life or when both testicles have descended into scrotum. Castration is not necessary for lambs to be slaughtered before six months of age, as long as they are separated from ewes before they are four months old.
- Protect pasture-born lambs from predators with fencing and/or guardian animals.
- Market winter-born lambs for spring holidays.
- Shear sheep as soon as weather permits, preferably before fly season.

- _____
- _____

PETS

- Brush dogs regularly during shedding season.
- Scrub litter boxes and set out to dry in the sun.
- Have pets tested for internal parasites and treat as necessary.
- Trim nails.

- _____
- _____
- _____

BUILDINGS & GROUNDS

- To minimize weed growth in lawns, spread sugar-ground corn gluten in grass when forsythia are blooming. The corn gluten acts as a natural pre-emergent herbicide for weeds.
- Sharpen and balance mower blades if not done yet.

- For best health, lawn should be 2 ½ - 3" tall and have at least 5% clover, which fixes nitrogen in the soil and "feeds" the grass.
- Clip or mow overgrown ground cover.
- Weed and edge perennial beds. Remove excess winter mulch.
- Plant trees and shrubs.
- Plant roses and daylilies.
- Finish pruning established roses. Cut out brown or crossed branches and prune for shape and size.
- Prune trellised vines.
- Fertilize bulbs as flowers fade.
- Make a list of desired bulbs for fall planting.

- _____
- _____
- _____

Ponds

- Put out hardy water lilies.

- _____
- _____

MACHINERY & EQUIPMENT

- Check fluid levels and tire pressure according to owners' manuals.
- Thoroughly wash and detail cars to remove last traces of road salt.
- Clean, lubricate and store snow plowing equipment.
- Purchase backup supplies of auto fluids.

- _____
- _____
- _____

ENJOYMENT

- If you enjoy fishing, buy license and prepare tackle and rods. Buy any needed supplies. Visit your favorite fishing spot or try out a new one.
- Create a relaxing spot to sit in the yard by installing a swing, picnic table or bench.
- Put out hummingbird feeders in mid-April.

- _____
- _____
- _____

5

MAY

Planting Time

May fulfills the promise of spring and offers the first delicious taste of summer. It also brings us back into the full swing of outdoor work on the homestead. By the end of the month most of us can plant our summer crops with no fear of frost damage. As we're completing our major planting in the garden, our livestock populations are probably at their highest point of the year and suddenly there is grass to mow every week. Amid the rush of spring work, take time out on Memorial Day to remember those who have made our freedom possible.

HOUSEHOLD

- Wash outside of house if necessary to remove dirt, algae and mildew.
- Clean porches.
- Wash and put away winter coats, boots and accessories.
- Clean and organize mud room.
- Clean out fireplaces and wood burners. Clean and store fireplace tools. Return unused wood to outdoor woodpile.
- Test smoke and CO detectors according to manufacturers' instructions.
- Clean rugs and carpets.
- Clean freezer and refrigerator and dispose of outdated food.
- Begin to use up last year's stored food. Clean root cellar and dispose of inedible produce.
- _____
- _____
- _____

GARDENING

- Plant corn after soil has warmed to at least 60 degrees.
- All summer vegetables can be planted by the end of the month.
- _____
- _____
- _____

HARVESTING/PRESERVING

- Harvest rhubarb; freeze, can, make pie, cake or preserves.
- Harvest asparagus; freeze or can.

MAY 51

- Harvest radishes and green onions.
- Pick lettuce, spinach and other spring greens.

- _____
- _____
- _____

FORAGING

- **WARNING: Many wild plants or parts of plants can be toxic! Before consuming any wild foods, be sure you can identify exactly what you are harvesting and that you know how to properly prepare it. Harvest only from clean sites not contaminated with chemical sprays, traffic pollution, animal waste or other toxins. Do not consume large amounts of any wild plant unless you are absolutely sure of its safety.**
- **Before harvesting and consuming wild plant foods, consult an experienced local forager or a reputable foraging field guide such as one of the following:** *A Field Guide to Edible Wild Plants: Eastern and Central North America* **by Lee Allen Petersen or** *The Complete Guide to Edible Wild Plants* **by the Department of the Army.**

- Pick dandelion greens and flower buds; use fresh or cooked.
- Pick chickweed for salads.
- The young leaves of plantain, lamb's quarters and garlic mustard can be eaten fresh or cooked.
- Pick leaves, young stems and roots of redroot pigweed. Use leaves and stems as fresh or cooked vegetable; cook and mash roots as you would potatoes.
- Bulrush shoots can be eaten raw or cooked.
- Pick milkweed shoots when they are 3"- 6" tall; boil 20 minutes in lightly salted water and serve with your favorite seasoning.
- Harvest the small bean-like seeds from the "helicopters" of silver

and red maples. Eat raw or cooked.
- Daylily buds and young green shoots can be eaten raw or lightly cooked, and the flowers can be eaten raw in salads, used as edible garnishes, added to soups and other cooked dishes, or dipped in batter and fried.
- Harvest young mint leaves; use fresh or dried.
- Harvest chamomile flowers; dry for tea. **(CAUTION: chamomile should not be used by pregnant women, as it is a uterine stimulant).**

- _____
- _____
- _____

LIVESTOCK

- Decrease or eliminate hay feedings as full grazing resumes.
- Mow pastures if grass is more than 8-10" tall, is about to go to seed, or if tall weeds become prevalent. Allow grass to remain at least 3-4" tall after mowing or grazing.
- Check fences and gates; repair as needed.
- Put up fly papers and traps.

- _____
- _____
- _____

Bees

- Requeen if necessary.
- Add supers as needed.
- Take swarm control measures as needed. Split large colonies.
- _____
- _____

Cattle

- If you have cows ready to calve, check calving supplies. Include a halter and lead rope for the cow, obstetrical gloves and lubricant, calving chains, towels, dull scissors and dental floss for cutting and tying off umbilical cord if necessary, iodine spray or iodine and dipping cup for navel, nasal aspirator syringe, veterinary thermometer, flashlight, ear tags if using, bottle and lamb nipple. Have a clean, dry stall available in case it's needed to warm calf.
- Study calving information if you are inexperienced or need a refresher.
- Be careful of your personal safety when approaching or handling cows that are calving or that have calves. Cows that need assistance with calving should be placed in a calving chute for the handler's safety. A properly constructed chute includes a head gate that restrains the cow while allowing her enough freedom to stand or lie down comfortably. The chute should have adjustable side panels that allow the handler or veterinarian safe access to the cow as well as allowing the calf to nurse. Plans for simple homemade chutes are available from many county Extension offices.
- Castrate bull calves when both testicles have descended into scrotum. Dehorn calves if desired.
- Breed now for February calving. Rebreed cows about three months after calving. Be sure cattle are healthy and fit for breeding.

- _____
- _____

Goats

- Check hooves; trim if needed. Avoid stressing heavily pregnant does with hoof trimming.
- When kidding season is over, clean, sanitize, organize and restock kidding kit.

- _____
- _____

Horses

- Trim hooves.
- Continue to adjust feed rations as grazing resumes and activity increases.

- _____
- _____

Pigs

- Purchase feeder pigs if needed. Look for healthy pigs that are 6-8 weeks old and weigh 40-60 pounds. Pigs should already have been wormed, vaccinated and castrated. Avoid unhealthy pigs, runts and uncastrated males. FYI: Gilts (female pigs) produce leaner meat than barrows (castrated males).
- Begin rotational grazing if raising pigs on pasture.

- _____
- _____

Poultry

- Purchase turkey poults late in the month to reach harvest weight by Thanksgiving. Keep turkeys isolated from other poultry to protect them from diseases.
- Scrub and sun-dry feeders and waterers.
- Thoroughly clean and disinfect brooders, incubators and all chick-raising equipment after final use.
- Begin pasturing chickens if not done in April. Take measures to protect birds from predators including dogs, foxes, coyotes, raccoons and hawks. The safest method is to keep birds enclosed in a "chicken tractor," or portable pen which you move to fresh grass every few days.
- Butcher ducks at 8 weeks of age, as soon as they are fully feathered and have no pinfeathers visible. Once birds begin to molt they will be difficult to pluck until their adult plumage is completely in.
- Expose geese bought as weeders to their intended forage before they are six weeks old. Pick the weeds you want them to eat and offer them as part of the goslings' diet before you turn them out to the field or garden.

- _____
- _____

Rabbits

- Clip nails on breeding stock.
- Separate male and female young by three months of age.
- If raising rabbits for meat, butcher young between eight weeks and eight months of age.
- Breed does nine months of age or older. On the 28th day after breeding, place nest boxes in cages of pregnant does.

- _____
- _____

Sheep

- Check hooves; trim if needed. Avoid stressing heavily pregnant ewes with hoof trimming.
- Dock lambs' tails at 2-3 days of age. Castrate males within the first week of life or when both testicles have descended into scrotum. Castration is not necessary for lambs to be slaughtered before six months of age, as long as they are separated from ewes before they are four months old. Complete all docking and castrating before fly season begins.
- Protect pasture-born lambs from predators with fencing and/or guardian animals.
- When lambing season is over, clean, sanitize, organize and restock lambing kit.
- Shear sheep as soon as weather permits, preferably before fly season begins.
- _____
- _____

PETS

- Trim nails.
- Watch for ticks. If fleas are a problem, vacuum carpets frequently and wash pet bedding weekly.
- _____
- _____
- _____

MAY 57

BUILDINGS & GROUNDS

- Regrade driveways and reset stepping stones in sidewalks if necessary.
- Consider using natural weed-control methods on sidewalks and driveways. The results won't be as dramatic as with chemical treatments, but they will be less toxic to the environment – and to you. Try spraying weeds with vinegar, pouring boiling salted water on them or covering them with clear or black plastic for 4-10 weeks, depending on temperature.
- Fill outdoor planters and hanging baskets; plant annuals late in the month.
- Plant perennials, summer bulbs and tubers, including gladiolus and dahlias.
- Prune needle evergreens.
- Sprinkle 1/3 to ½ cup Epsom salts around rose bushes; fertilize with aged manure when new leaves appear.
- Deadhead bulbs as blossoms fade.
- Finish spring cleanup of fallen leaves, dead plant material and fallen branches and twigs.
- Mulch newly planted trees and shrubs.
- Clean and organize greenhouse.
- Dig and divide perennials after blooming.
- Pinch back chrysanthemums.
- Prune flowering shrubs after blooms have faded.

- _____
- _____
- _____

Ponds

- Return fish and tender plants to small ponds.

- _____
- _____

MACHINERY & EQUIPMENT

- Check fluid levels and tire pressure according to owners' manuals.
- Remove winter tires.
- _____
- _____
- _____

ENJOYMENT

- Look for monarch butterfly eggs on the underside of milkweed leaves. They look like tiny yellowish-green cones and are laid singly. After three to five days, the eggs will hatch into boldly marked zebra-striped caterpillars.
- Clean and put away bird feeders.
- Put out hummingbird feeders if not done in April.
- Fill bird baths. Change water every 2-3 days and scrub off algae and dirt weekly.
- Enjoy a Memorial Day picnic and remember those who have made our freedom possible.
- _____
- _____
- _____

6

JUNE

Welcoming Summer

June is one of the most enjoyable months on the homestead. The weather is pleasant, the crops are producing and the animals are grazing on thick pastures. If you have children, they'll be out of school and available to help with the chores. Summer is here and the living may not be easy, but in spite of all the work the season brings, there is much to appreciate on the homestead!

HOUSEHOLD

- Clean dust from fans and ceiling fans before using them for the first time.
- Test smoke and CO detectors according to manufacturers' instructions.
- Empty and clean dehumidifiers frequently to prevent mold growth.
- Clean and organize laundry area.
- Remove bottom rack of dishwasher and inspect drain area for debris. Clean drain area and rubber door gaskets with soapy cloth or small brush. Place a cup of white vinegar on the top rack of empty dishwasher, then run the machine on the hottest cycle to remove mineral deposits and odors.
- Soak shower heads in vinegar to remove mineral deposits.
- Purchase needed equipment and supplies for canning and freezing produce.
- _____
- _____
- _____

GARDENING

- Start seeds of broccoli, Brussels sprouts and cabbage indoors for fall crops.
- Plant lettuce, rutabagas, pumpkins and winter squash in the garden.
- As weather warms, consider planting bolt-resistant varieties of lettuce, such as Red Sails, summer crisps (Batavian lettuce) or Green Forest romaine. Plant lettuce every two to three weeks for continuous harvest.
- Put floating row cover over plants vulnerable to insects, rabbits or deer.
- Turn compost.

JUNE

- Water gardens in the mornings for best plant health. This allows leaves to dry off before nightfall, helping to prevent fungal diseases.
- Pinch back herbs.
- Mulch vegetables with grass clippings, straw, black plastic, compost or other material. A thick layer of wet newspapers or cardboard under organic mulches will help suppress weeds and conserve moisture.
- Put bird netting or other barriers over berry bushes.
- Carefully cut suckers and water sprouts from fruit trees. These are the thin, soft branches that shoot straight up from the base and trunk of the tree.

- _____
- _____
- _____

HARVESTING/PRESERVING

- Check ripening fruits and vegetables daily and pick them as they reach their peak, rather than letting them become over-mature.
- Pick strawberries. Dry, freeze or make jam.
- Pick red raspberries. Freeze or make jam.
- Pick and dry lavender.
- Pick cherries; freeze, can or make pies.
- Pick lettuce, spinach and other spring greens.
- Harvest peas, beets, turnips, broccoli, cauliflower, cabbage, Swiss chard and other vegetables as they become ready.

- _____
- _____
- _____

FORAGING

- **WARNING: Many wild plants or parts of plants can be toxic! Before consuming any wild foods, be sure you know exactly what you are harvesting and how to properly prepare it. Harvest only from clean sites not contaminated with chemical sprays, traffic pollution, animal waste or other toxins. Do not consume large amounts of any wild plant unless you are absolutely sure of its safety.**
- **Before harvesting and consuming wild plant foods, consult an experienced local forager or a reputable foraging field guide such as one of the following:** *A Field Guide to Edible Wild Plants: Eastern and Central North America* **by Lee Allen Petersen or** *The Complete Guide to Edible Wild Plants* **by the Department of the Army.**

- Begin to harvest purslane. Cook in a small amount of water or vegetable broth.
- Pick chickweed for salads.
- Harvest young leaves of plantain, lamb's quarters and garlic mustard. Use fresh or cooked.
- Pick leaves, young stems and roots of redroot pigweed. Use leaves and stems as fresh or cooked vegetable; cook and mash roots as you would potatoes.
- Pick wild strawberries. Use fresh, freeze or make jam.
- Harvest the small bean-like seeds from the "helicopters" of silver and red maples. Eat raw or cooked.
- Harvest milkweed flower bud clusters. Cook like broccoli or use in stir-fries and soups.
- Daylily buds and young green shoots can be eaten raw or lightly cooked, and the flowers can be eaten raw in salads, used as edible garnishes, added to soups and other cooked dishes, or dipped in batter and fried.
- Harvest young mint leaves; use fresh or dried.

- Harvest chamomile flowers; dry for tea. **(CAUTION: chamomile should not be used by pregnant women, as it is a uterine stimulant).**
- Dig chicory roots; dry and grind for coffee substitute.

- _____
- _____
- _____

LIVESTOCK

- Clean out hay storage areas. Put oldest hay to the front, where it will be used first. Use moldy or weedy hay for mulch, compost or other purposes.
- Make or buy hay. Be sure hay is thoroughly dry before stacking it in barn; damp hay can become hot enough to start a fire.
- Be sure animals have enough shade, fresh water and salt.
- Examine pastures for poisonous plants and trees.
- Check pastures for hazards such as groundhog holes and discarded trash.
- Keep grass and weeds trimmed around livestock buildings and under electric fences.
- Check fences and gates; repair as needed.
- Check pastures after storms for fallen branches and damaged fences.
- Rotate pastures as needed.
- Continue or step up fly control program.
- Have halters and ropes available to lead animals out of barn in case of fire, as well as hoods or blindfolds for panicked animals. Keep an axe in the barn in case it is needed to break doors open. Have a fire extinguisher at each end of the barn and check them once a year. Be sure exit doors are not blocked. Plan and practice barn fire drill. Ask neighbors to participate, in case they are the only ones around when a fire breaks out. Decide where you will take animals once they are removed from the barn.

- _____
- _____
- _____

Bees

- Harvest honey.
- Take swarm control measures as needed.

- _____
- _____

Cattle

- When calving season is over, clean, sanitize, organize and restock calving supplies.
- Breed now for March calving. Rebreed cows about three months after calving. Heifers should be at least 15 months old before first breeding.
- Use excess milk to make butter, yogurt, ice cream or cheese.
- Be sure lactating cows have plenty of water. A dairy cow will drink 20-40 gallons per day during the summer.

- _____
- _____

Goats

- Freeze excess milk or make cheese, yogurt, fudge, ice cream or soap.

- _____
- _____

Horses

- Begin bot control program. Remove eggs from horses' legs with bot block or safety razor.
- Monitor body condition to be sure horses are not getting too fat on lush grass.

- _____
- _____

Pigs

- Purchase feeder pigs if not done by now. Prices are often lowest in early summer, when supply is traditionally high.
- Pigs can benefit from discarded garden produce and excess cow or goat milk as supplements to their regular feed.
- If you are raising pigs for income, consider marketing roasting pigs for summer barbecues.
- Breed sows now for farrowing in September or October. Estimate farrowing date on calendar at three months, three weeks and three days after breeding date.

- _____
- _____

Poultry

- Purchase turkey poults early in the month if not done in May. Keep turkeys isolated from other poultry to protect them from diseases.
- Butcher broiler chickens at 8-12 weeks of age.

- _____
- _____

Rabbits

- Examine litters of newborn bunnies; remove dead ones and place excess or rejected ones with foster doe if available.

- _____
- _____

Sheep

- _____
- _____

PETS

- Trim nails.
- Watch for ticks and fleas. Treat flea infestations promptly to keep them from getting out of hand.

- _____
- _____
- _____

BUILDINGS & GROUNDS

- Pressure-wash, repair and stain or paint decorative fences, gates and decks.
- Prune forsythia, lilac, quince, rhododendron and wisteria after blooms fade.
- Finish renovating perennial beds and moving plants and bulbs.
- Remove suckers and water sprouts from landscape trees.

JUNE **67**

- Fertilize annuals and roses.
- Scatter coffee grounds under rhododendrons, hollies and azaleas to maintain acidity levels. Refresh mulch if necessary, using shredded pine bark, oak leaf mold or pine needles.
- Keep flower beds edged and groomed.
- Pinch back chrysanthemums.
- Deadhead roses, perennials and annuals.
- Remove or spray poison ivy vines carefully. Very small vines can be removed by putting a bread bag or newspaper bag over one's hand, pulling out the vine with it, and then pulling the bag up over the vine to dispose of it.

- _____
- _____
- _____

Ponds

- _____
- _____

MACHINERY & EQUIPMENT

- Check vehicle fluids and tire pressure according to owners' manuals.
- Perform or schedule routine maintenance of tractors and other equipment according to manufacturers' directions.
- Do not allow children to ride on or operate farm machinery. See www.farmsafetyforjustkids.org for more information.

- _____
- _____
- _____

ENJOYMENT

- Try an old-fashioned picnic on the lawn, complete with checkered tablecloth or blanket.
- If you have kids or grandkids, celebrate the last day of school with a special treat or activity.
- Set up a badminton or volleyball net for impromptu games in spare moments.

- _____
- _____
- _____

7
JULY

Celebrating Independence

This month we celebrate our independence – not just as a country, but also as homesteading individuals. By providing for ourselves, we can be free in so many ways: free from reliance on grocery stores, free from worries that our food might be contaminated or genetically modified, and free from undue dependence on utility providers and the consumer economy. As homesteaders, we can spend less time worrying about what the government is doing for us and more time concentrating on what we are doing for ourselves – just as our founding fathers did when they declared our country's independence.

HOUSEHOLD

- Practice household fire drill. Plan at least two escape routes from each room. Make sure windows and doors open fully and are unobstructed. If your house has more than one story, have a collapsible fire ladder in an accessible location on each upper level. Decide on an outdoor meeting place in the event of a fire.
- Take photos of house contents and store in fireproof safe with important papers.
- Find and destroy wasp and hornet nests. Look for them under eaves and porch roofs, in attics, near outdoor light fixtures, in children's play structures and other sheltered areas.
- Clean filters on window air conditioners.
- Check furnace filter if using central air; clean or change as needed.
- Test smoke and CO detectors according to manufacturers' instructions.
- If you have a spring or well, have water tested to be sure it is free of contaminants.
- Pour a cup or so of water into basement floor drains and unused sink drains to block sewer gases from backing up into house.
- Clean and organize basement – it's a cool place to work on a hot day.

- _____
- _____
- _____

GARDENING

- Pick, weed, water, repeat!
- Plant beets, cucumbers, zucchini and turnips.
- Continue planting other summer vegetables as desired. Make a second planting of green beans for continuous harvest.
- Toward end of month, plant seedlings of cabbage, broccoli and cauliflower for fall harvest.

- Keep compost pile moist.
- Stake, prune and mulch tomatoes.
- Provide afternoon shade for salad greens by propping a section of lattice or a pallet over the bed.

- _____
- _____
- _____

HARVESTING/PRESERVING

- Cut and dry herbs.
- Continue to harvest and preserve summer vegetables.
- Begin canning beets, pickles and tomatoes. Setting up a canning area centered around a spare stove in the basement will keep the heat and mess out of the kitchen.
- Make tomato paste and freeze in ice cube trays for convenient single-serving portions.
- Corn may be ready for harvest late in the month.
- Carefully dig "new" potatoes when plants begin to flower.
- Pick raspberries and blueberries; use fresh, freeze or make jam.
- Pull and dry garlic when the bottom leaves are brown but the top few are still green. Clean off excess dirt and trim roots to about ¼." Allow garlic to cure by spreading the plants out in a dry, covered area (such as a porch, ventilated shed or garage) for two weeks, or until necks and roots are thoroughly dry. Set aside the best bulbs for fall planting. Braid or cut off leaves from remaining bulbs and store in a cool, well-ventilated area.
- Harvest July apples.
- Pick gooseberries.

- _____
- _____
- _____

FORAGING

- **WARNING:** Many wild plants or parts of plants can be toxic! Before consuming any wild foods, be sure you know exactly what you are harvesting and how to properly prepare it. Harvest only from clean sites not contaminated with chemical sprays, traffic pollution, animal waste or other toxins. Do not consume large amounts of any wild plant unless you are absolutely sure of its safety.
- Before harvesting and consuming wild plant foods, consult an experienced local forager or a reputable foraging field guide such as one of the following: *A Field Guide to Edible Wild Plants: Eastern and Central North America* **by Lee Allen Petersen or** *The Complete Guide to Edible Wild Plants* **by the Department of the Army.**

- Harvest purslane. Cook in a small amount of water or vegetable broth.
- Pick chickweed for salads.
- The tender top leaves of lamb's quarters can be used fresh or cooked.
- Pick young plantain leaves and seed heads. Use fresh or cooked.
- Pick young leaves, upper stems and roots of redroot pigweed. Use leaves and stems as fresh or cooked vegetable; cook and mash roots as you would potatoes.
- Harvest milkweed flower bud clusters. Cook and eat like broccoli or use in stir-fries or soups. Pick milkweed pods while they are still small (1-3") and silk is tender and juicy. Cook like okra or use in stir-fries and soups.
- Daylily buds can be eaten raw or lightly cooked like green beans, and the flowers can be eaten raw in salads, used as edible garnishes, added to soups and other cooked dishes, or dipped in batter and fried.
- Harvest young mint leaves and use fresh or dried.
- Wild blueberries and huckleberries are beginning to ripen. Use

fresh, freeze or make jelly.
- Pick wild raspberries and blackberries. Use fresh, freeze or make jelly.
- Harvest ground cherries (husk tomatoes) when fruits and husks have turned yellow and dropped off the plant. Eat fresh, freeze or make into jams or pies.

- _____
- _____
- _____

LIVESTOCK

- Monitor natural water sources for quantity and quality.
- Be careful not to overstress animals on hot days.
- Be sure pastured animals have adequate shade, water and salt.
- Rotate pastures as needed.
- Check fences and gates; repair as needed.
- Mow pastures if grass is more than 8-10" tall, is about to go to seed, or if tall weeds become prevalent. Allow grass to remain at least 3-4" tall after mowing or grazing.
- Control flies and rodents.
- Monitor and treat for external parasites as needed.
- Be sure that any animals kept indoors have adequate ventilation and water.
- Update written animal care plan for others to use in case you are away from home in an emergency or are otherwise unable to care for your animals. Post copies of the plan in a central area of the barn or other animal housing, in feed storage areas and in the house. Notify designated emergency caretakers of any changes in the routine.

- _____
- _____
- _____

Bees

- Harvest and sell honey.
- Inspect for mites. Treat if necessary

- _____
- _____

Cattle

- Breed now for April calving. Rebreed cows about three months after calving. Heifers should be at least 15 months old before first breeding.
- Be sure lactating cows have plenty of water. A dairy cow will drink 20-40 gallons per day during the summer months.
- Use excess milk to make butter, ice cream, yogurt or cheese.

- _____
- _____

Goats

- Check hooves; trim if needed.

- _____
- _____

Horses

- Trim hooves.
- Remove bot eggs with bot block or safety razor.

- _____
- _____

Pigs

- Make sure pigs have plenty of clean water and shade. A wallow will help keep pigs cool and happy, but they need a separate source of fresh water for drinking.
- Pigs can benefit from discarded garden produce and excess cow or goat milk as supplements to their regular feed.
- If you are raising pigs for income, summer is the time to market roasting pigs.
- Breed sows now for farrowing in October or November. Estimate farrowing date on calendar at three months, three weeks and three days after breeding date.

- _____
- _____

Poultry

- Be sure that confined birds have adequate ventilation and water.
- Scrub and sun-dry feeders and waterers.

- _____
- _____

Rabbits

- Remove young from doe's hutch at six weeks of age.
- Clean cages, hutches and nest boxes. Scrub feeders and waterers.
- Be sure rabbit hutches have adequate ventilation on hot days.

- _____
- _____

Sheep

- Check hooves; trim if needed.

- _____
- _____

PETS

- Trim nails.
- Continue flea and tick treatments.

- _____
- _____
- _____

BUILDINGS & GROUNDS

- Keep weeds and grass trimmed around outbuildings to reduce fire hazards and limit rodent hiding places.
- Trim hedges.
- Fertilize chrysanthemums, annuals and container plants.
- Order bulbs and trees for fall planting.
- Deadhead flowers.
- Pinch back chrysanthemums one last time before July 4th.
- Keep flower beds weeded and mulched.
- Cut back wisteria severely after bloom.
- Eliminate sources of standing water to reduce mosquito population.

- _____
- _____
- _____

Ponds

- Be sure that small ponds have sufficient aeration for fish health.

- _____
- _____

MACHINERY & EQUIPMENT

- Check vehicle fluids and tire pressure according to owners' manuals.
- Keep mower decks and string trimmers free of caked grass clippings.

- _____
- _____
- _____

ENJOYMENT

- Host an Independence Day campfire cookout. For true independence, incorporate as many home-grown and foraged foods as possible.
- Make homemade ice cream. If you don't have an ice cream machine, go online to find instructions for making a simple one out of two coffee cans.

- _____
- _____
- _____

8

AUGUST

Reaping the Rewards

The stifling heat of late summer can make it difficult to work outdoors this month. Our gardens, however, are at the height of production and require our care whether we feel like it or not. To reap the benefits of all that we have done so far, we have to keep watering, weeding and harvesting throughout the hottest days of the year. Water in the mornings to allow plants to dry off by evening, thus helping to reduce fungal diseases. Morning is also the best time to harvest produce, while quality is at its peak. Weeding and planting can be done in the evening. If you must work in the sun, wear a broad-brimmed hat and drink plenty of water. Be sure your animals have respite from the sun as well. In August, shade and water are the best friends of man and beast.

HOUSEHOLD

- Order heating fuel now to take advantage of off-season prices.
- Clean and organize refrigerator and freezer. Discard outdated food.
- Check basement, attic and pantry for insect infestations; treat as needed.
- Check furnace filter if using central air; clean or change as needed.
- Test smoke and CO detectors according to manufacturers' instructions.
- Clean filters on window air conditioners.
- Keep dehumidifiers clean to avoid mold formation.
- Build or prepare root cellar for winter storage of fruits and vegetables.

- _____
- _____
- _____

GARDENING

- Keep gardens mulched to preserve moisture and suppress weeds.
- Plant out broccoli, cabbage and Brussels sprouts seedlings if not done in July.
- Plant turnips and carrots for fall and winter harvest.
- Keep melons well watered for best fruit development.
- Late in the month, pinch back tomato plants to concentrate energy in existing fruits.
- Prop sections of lattice or other shade structures over lettuce beds, or plant lettuce where it will receive afternoon shade.
- Toward the end of the month, make a third planting of green beans if desired.
- Remove suckers from fruit trees if not done earlier.

- _____
- _____
- _____

HARVESTING/PRESERVING

- Harvest onions about ten days after most of the tops have turned yellow or brown and fallen over. Allow to dry in the sun for a day or so and then cure by spreading them out in a covered area (such as a porch, ventilated shed or garage) for two weeks, or until necks and outer skins are thoroughly dry. Then cut off leaves at least 1" above neck. Store in a ventilated container in a cool, dry, dark area.
- Freeze or can tomatoes, corn, beans, zucchini, peppers and other vegetables. Canning can be more pleasant if done early in the day, with fans or air conditioning running, or on a spare stove in the basement.
- Pick and dry herbs.
- Harvest cucumbers; eat fresh or make pickles.
- Pick peaches. Can or freeze excess fruit.
- Pick pears when fruit is mature but still hard. Allow to ripen off the tree; eat fresh or can.
- Harvest blueberries and blackberries; eat fresh, freeze, dry or make jam or pies.

- _____
- _____
- _____

FORAGING

- **WARNING: Many wild plants or parts of plants can be toxic! Before consuming any wild foods, be sure you know exactly what you are harvesting and how to properly prepare it. Harvest only from clean sites not contaminated with chemical sprays, traffic pollution, animal waste or other toxins. Do not consume large amounts of any wild plant unless you are absolutely sure of its safety.**
- **Before harvesting and consuming wild plant foods, consult an experienced local forager or a reputable foraging field guide such as one of the following:** *A Field Guide to Edible Wild Plants: Eastern and Central North America* **by Lee Allen Petersen or** *The Complete Guide to Edible Wild Plants* **by the Department of the Army.**

- Harvest purslane; cook in a small amount of water or vegetable broth.
- Pick chickweed for salads.
- Harvest tender top leaves of lamb's quarters; eat fresh or cooked.
- Pick young leaves, upper stems and roots of redroot pigweed. Use leaves and stems as fresh or cooked vegetable; cook and mash roots as you would potatoes.
- Harvest milkweed flower bud clusters. Cook and eat like broccoli or use in stir-fries and soups. Pick milkweed pods while they are still small (1-3") and silk is tender and juicy. Cook like okra or use in stir-fries and soups.
- Daylily buds can be eaten raw or lightly cooked like green beans, and the flowers can be eaten raw in salads, used as edible garnishes, added to soups and other cooked dishes, or dipped in batter and fried.
- Pick huckleberries and blackberries; eat fresh or preserve by drying, freezing or making jam, jelly or pie filling.
- Harvest ground cherries (husk tomatoes) when fruits and husks have turned yellow and dropped off the plant. Eat the fruits fresh,

freeze them or make jams or pies.
- Pick elderberries. Make jam, juice or wine; freeze juice in ice cube trays to have on hand for its antiviral properties.
- Harvest crabapples for cider, apple butter, spiced apples or jelly.

- _____
- _____
- _____

LIVESTOCK

- Purchase or make second cutting of hay.
- Check pastures for hazards such as poisonous plants and trees, groundhog holes, trash and fallen branches.
- Check fences and gates; make repairs as needed.
- Check natural water supplies for quality and quantity. Be sure automatic waterers are working properly and that water troughs and buckets are kept clean and filled.
- Make sure animals have shelter from sun and flies.
- Be careful not to overstress animals on hot days.
- Schedule fall vaccinations.

- _____
- _____
- _____

Bees

- Requeen if needed.
- Combine and feed weak hives.

- _____
- _____

Cattle

- Breed now for May calving. Rebreed cows about three months after calving. Heifers should be at least 15 months old before first breeding.
- Trim hooves on dairy cows if needed.
- Be sure lactating cows have plenty of water. A dairy cow will drink 20-40 gallons per day during summer.
- _____
- _____

Goats

- Plan fall breedings and choose buck.
- _____
- _____

Horses

- _____
- _____

Pigs

- Provide adequate shade and/or wallows to keep pigs cool on hot days. Be sure pigs also have access to clean drinking water.
- Pigs can benefit from discarded garden produce and excess cow or goat milk as supplements to their regular feed.
- Butcher hogs when they reach about 250 pounds.
- If you are raising pigs for income, market roasting pigs for summer barbecues.

AUGUST

- _____
- _____

Poultry

- Provide adequate ventilation and water in coops.

- _____
- _____

Rabbits

- Be sure rabbits have adequate ventilation and fresh water.
- Clip nails on breeding stock.
- Separate male and female young before they are three months old.
- If raising rabbits for meat, butcher between eight weeks and eight months of age.
- Breed does nine months of age or older. On the 28th day after breeding, place nest boxes in cages of pregnant does.

- _____
- _____

Sheep

- Breed now to have January-born lambs for Easter. Estimate due date on calendar, about 145 days after breeding.
- Shear lambs 4-6 weeks before butchering in order to use or sell pelts as shearling.

- _____
- _____

PETS

- Bathe and groom dogs.
- Wash litter boxes and allow to dry in the sun.
- Watch for ticks.
- Continue flea control program.
- Trim nails.

- _____
- _____
- _____

BUILDINGS & GROUNDS

- Order or purchase trees and bulbs for fall planting.
- Continue to deadhead annuals and perennials.
- Lawns may go dormant and turn brown in dry weather. This is natural – do not try to "save" the grass by fertilizing it. Simply let it rest until cool, wet weather returns in the fall.
- Remove sources of standing water to minimize mosquito populations.

- _____
- _____
- _____

Ponds

- Make sure small ponds have adequate aeration for fish health.

- _____
- _____

MACHINERY & EQUIPMENT

- Check vehicle fluids and tire pressure according to owners' manuals.
- Add air to wheelbarrow, cart and trailer tires if needed.

- _____
- _____
- _____

ENJOYMENT

- Host a back-to-school party for the kids.
- Have a summer "garden party" featuring roasted or boiled corn on the cob, fresh tomato salad, grilled or stuffed zucchini and other homegrown goodies.
- Take a few minutes during the day and night to listen to the insect chorus. How many different songs can you hear?

- _____
- _____
- _____

9

SEPTEMBER

Harvesting Autumn's Bounty

September may feel like an extra month of summer, but amid its warmth and sunshine our attention gradually turns from planting and propagating the garden to harvesting its bounty. Winter squash, turnips, potatoes, cabbage and other traditional fall and winter foods are ready to pick and cure for long-term storage. Summer vegetables are still producing, and fall fruits like apples and Asian pears are ripening. Careful preservation will hold these treasures to be enjoyed throughout the winter and even into early spring.

HOUSEHOLD

- Have furnaces, fireplaces, wood burners and chimneys cleaned and inspected. Check for cracks, loose bricks, separations in flue pipes, heat damage to shingles around chimney, and cracks in exterior chimney. Have furnace serviced. Check or install chimney caps to keep animals and birds out. Remember to open dampers before using fireplaces for the first time.
- Check heating fuel supplies.
- Check furnace filter if using central air; clean or change as needed.
- Test smoke and CO detectors according to manufacturers' instructions.
- Step up rodent control measures as mice and squirrels begin to scout out homes for winter. Block off openings in foundation, attic and around windows and doors. Scatter dried mint leaves in storage areas to help repel mice.
- Remove, clean and store window air conditioners when cooling season is over.
- Reseason cast iron skillets. Lay a sheet of aluminum foil on the center oven rack and preheat oven to 350 degrees. Wipe the skillet with a small amount of vegetable oil or lard. Wipe off excess with a paper towel. Place the skillet upside down on the aluminum foil. Heat for an hour and let cool gradually. Be prepared to ventilate the kitchen in case the seasoning process produces smoke. To maintain the nonstick surface, avoid boiling water or using acidic liquids in the skillet. Wipe the skillet clean immediately after every use – do not soak in water.
- Mix a fresh batch of household potting soil. Try a mixture of 1-2 parts soil, 1 part builder's sand or perlite and 1 part peat moss, compost or leaf mold. Add 1 tablespoon bone meal per quart of mix.
- Repot houseplants as needed.
- _____
- _____
- _____

GARDENING

- Plant lettuce, spinach, endive and other greens for fall and early winter use. Good greens for late-season planting include spinach, arugula, kale, romaine, 'Winter Marvel' looseleaf lettuce and 'Arctic King' or 'North Pole' butterhead.
- Plant corn salad (mache) early in the month for winter greens.
- Plant berry bushes and asparagus.
- Save non-hybrid tomato seeds by squeezing tomato contents into a glass of water. Stir once a day for three days. Drain and dry on a coffee filter or old t-shirt. When thoroughly dry, store in an airtight container with a packet of silica gel.
- Save seeds from other non-hybrid plants that you wish to grow next year. Dry thoroughly before storing in a cool, dry place.
- Build or repair cold frames as needed.

- _____
- _____
- _____

HARVESTING/PRESERVING

- Begin to harvest pumpkins, gourds and winter squash when they reach full color and skins are tough. Cure at 80-85 degrees and about 80% humidity for ten days, then store in a cool (50-55 degrees), dry place.
- Pick cabbage, broccoli, cauliflower, turnips and celery.
- Make sauerkraut.
- Pick and dry mature sunflower heads before birds clean them out. Store seeds in rodent-proof containers.
- Harvest apples and Asian pears as they ripen. Use damaged fruit quickly; store others in root cellar or preserve by drying or canning.
- Continue to harvest and preserve summer vegetables.
- Harvest grapes when they have a white bloom on skins and their seeds have turned brown. Make juice or wine, or dry for raisins.

- Harvest nuts as they begin to ripen. Pick nuts from the tree rather than waiting for them to drop if squirrels are a problem. Use fresh or store in freezer for several months.
- Dig up storage potatoes after plants have been brown for two weeks. Choose a dry, overcast day. Allow potatoes to dry for an hour or so before curing them for one or two weeks in a dark, cool (55-60 degrees) humid place. Store long-term in a dark, cold (35-40 degrees) moderately humid area with good ventilation.
- Dig sweet potatoes before the first frost. Choose a dry, overcast morning. Carefully dig the potatoes and brush off excess dirt. Let the potatoes sit out to dry for several hours, then place in shallow boxes and move to a warm (about 85 degrees), humid and well-ventilated place for a week or two to cure. Use any damaged potatoes right away or cook and freeze them; store others in root cellar or other cool (55-60 degrees) ventilated area with 75-80% humidity.

- _____
- _____
- _____

FORAGING

- **WARNING: Many wild plants or parts of plants can be toxic! Before consuming any wild foods, be sure you know exactly what you are harvesting and how to properly prepare it. Harvest only from clean sites not contaminated with chemical sprays, traffic pollution, animal waste or other toxins. Do not consume large amounts of any wild plant unless you are absolutely sure of its safety.**
- **Before harvesting and consuming wild plant foods, consult an experienced local forager or a reputable foraging field guide such as one of the following:** *A Field Guide to Edible Wild Plants: Eastern and Central North*

America **by Lee Allen Petersen or** *The Complete Guide to Edible Wild Plants* **by the Department of the Army.**

- Pick chickweed for salads.
- Harvest purslane; cook in a small amount of water or vegetable broth.
- Harvest new leaves of lamb's quarters. Eat fresh or cooked.
- Pick plantain leaves and seeds; use fresh or cooked.
- Pick young leaves, upper stems and roots of redroot pigweed. Use leaves and stems as fresh or cooked vegetable; cook and mash roots as you would potatoes.
- Pick milkweed pods while they are still small (1-3" long) and silk is tender and juicy. Cook like okra or use in stir-fries and soups.
- Daylily buds can be eaten raw or lightly cooked like green beans, and the flowers can be eaten raw in salads, used as edible garnishes, added to soups and other cooked dishes, or dipped in batter and fried.
- Harvest ground cherries (husk tomatoes) when fruits and husks have turned yellow and dropped off the plant. Eat fresh, freeze or make jams and pies.
- Harvest wild grapes. Eat fresh, make juice, jelly or dry for raisins.
- Harvest crab apples for cider, apple butter, spiced apples or jelly.
- Harvest seeds of dock, bulrush, plantain, redroot pigweed and lamb's quarters. Use as a grain or grind and use for flour and thickener. Redroot seeds can be "popped" like popcorn.
- Harvest the small bean-like seeds from "helicopters" of sugar maples. Eat raw or cooked.
- Harvest nuts as they ripen and begin to fall. These include chestnuts, black walnuts, shagbark hickory, beechnuts, butternuts and acorns. Use immediately or freeze for several months.
- Dig chicory roots; dry and grind for coffee substitute.
- _____
- _____
- _____

LIVESTOCK

- Mow pastures if grass is more than 8-10" tall, is about to go to seed, or if tall weeds become prevalent. Allow grass to remain at least 3-4" tall after mowing or grazing.
- Continue to rotate pastures.
- Check fences and gates; repair as needed.
- Reseed thinning pastures. Contact local Extension agent for seed recommendations for your area.
- Check quality and quantity of natural water sources used for livestock.
- Check hay and bedding supplies for winter.
- Sell or butcher unwanted animals before winter arrives.
- Arrange for butchering or sale of young animals that have reached market size. Generally, beef cattle are butchered at the end of their second summer, hogs at about 250 pounds, lambs at 90-100 pounds (about 5-6 months old) and goat kids at 6-8 months. However, animals can be slaughtered at whatever age or weight suits the owner's or buyer's personal preferences and needs.
- _____
- _____
- _____

Bees

- Harvest honey.
- Requeen if needed.
- Leave enough honey from fall harvest for bees, approximately 60 to 80 pounds depending on your climate. Feed 2:1 sugar/water solution if needed.
- Remove honey supers.
- Late in the month, install mouse guards/entrance reducers.

- _____
- _____

Cattle

-
-

Goats

- Record breeding dates and estimate due dates on calendar, 150 days later.
- Check hooves; trim if needed.

-
-

Horses

- Trim hooves.

-
-

Pigs

- Breed sows late in the month for January farrowing. Estimate farrowing date on calendar at three months, three weeks and three days after breeding date.
- Prepare farrowing stalls for sows bred in summer. Have a safely installed heat lamp or other source of warmth available for piglets.
- Prepare farrowing kit, including clean towels, iodine and a small cup for navel dipping, elbow-length plastic gloves, OB lubricant, flashlight, nasal aspirator syringe, colostrum and/or milk replacer or goat milk, feeding tube, human baby bottles and nipples.

- Be very careful of your personal safety around farrowing sows and those raising piglets. Even "pet" sows can become dangerously protective of their young.
- Dip navels of newborn piglets. Warm small, listless piglets under heat lamp if needed. Notch or tattoo ears if needed for identification. Be sure all piglets have a chance to nurse. Give iron shots or supplements to piglets at 3-5 days of age if they do not have access to soil in their pen. Clip tips of sharp canine teeth ("needle teeth") if desired. Castrate males at about one week of age.

- _____
- _____

Poultry

- Scrub and sun-dry feeders and waterers.
- Begin using deep-litter system for winter warmth. First, clean out poultry housing by removing old litter. Let buildings or coops air out for the day. Add 4-5" of fresh litter. Throughout fall and winter, add thin layers of fresh litter as needed until overall depth reaches 9-12". Remove wet or very dirty spots periodically as needed. Litter should remain just slightly moist and warm, but should not be wet or dusty and should not reek of ammonia.
- Cull unproductive hens. Healthy laying hens will have a moist, white, oval vent; a full red comb; full, soft wattles and earlobes; a full, soft abdomen and bright eyes. Hens with shriveled yellow vents, small abdomens and faded combs and wattles are usually not layers. To find out for sure, confine suspect hens individually for up to two weeks to see if they lay any eggs.

- _____
- _____

Rabbits

- Purchase new breeding stock if needed.
- Breed does that are at least nine months old. Breeding two or more does at once will allow for fostering of excess or rejected young.
- On the 28th day after breeding, place a nest box in each pregnant doe's cage.

- _____
- _____

Sheep

- Check hooves; trim if needed.
- Shear lambs 4-6 weeks before butchering in order to use or sell pelts as shearling.
- Breed now for February lambing. Estimate due date on calendar, about 145 days after breeding.

- _____
- _____

PETS

- Trim nails.
- Continue to check for ticks and fleas.
- Test and treat for internal parasites.
- Wash bedding.

- _____
- _____
- _____

BUILDINGS & GROUNDS

- Do any needed exterior painting.
- Labor Day weekend is a good time to plant grass seed, including overseeding thinning lawns.
- Plant early bulbs including crocus and snowdrops.
- Water bulb beds and newly planted trees if season is dry.
- Edge, weed and mulch flower beds.
- Do major landscape renovations.
- Plant chrysanthemums, peonies and wildflower seeds.
- Dig and divide perennials.

- _____
- _____
- _____

Ponds

- _____
- _____

MACHINERY & EQUIPMENT

- Check fluids and tire pressure in vehicles.

- _____
- _____
- _____

ENJOYMENT

- Although Labor Day marks the unofficial end of summer, the weather is still quite warm in many areas. Now is the time to get in the last of those fun summer activities, before autumn weather arrives.

- _____
- _____
- _____

10
OCTOBER

Preparing for Winter

While the sunny days and clear skies of October beckon us to soak up the last of summer's warmth, it's important to focus this month on preparing our homestead for winter. This means winterizing the house and livestock buildings, laying in stores of food for ourselves and our animals, cleaning up the gardens and preparing our vehicles for snowy roads. Take advantage of good weather to get things in order before outdoor work becomes unpleasantly chilling. If it all seems overwhelming, just remember to take it one step at a time and know that each completed task brings you that much closer to greeting winter weather with confidence.

HOUSEHOLD

- Finish preparing heating systems for winter use, including cleaning chimneys. Check for cracks, loose bricks, separations in flue pipes, heat damage to shingles around chimney, and cracks in exterior chimney. Have furnace serviced. Check or install chimney caps to keep animals and birds out. Remember to open dampers before using fireplaces for the first time.
- Check heating fuel supplies; order more if needed.
- Invest in a programmable thermostat. Set it to reduce the temperature during the night as well as during the day if everyone will be away from home.
- Be sure firewood is kept dry and covered.
- Test smoke and CO detectors according to manufacturers' instructions.
- Take out window screens, clean windows and replace storm windows.
- Fill in gaps around windows and doors with caulk, expanding foam insulation or weatherstripping.
- Look for gaps where wiring, plumbing and other utilities enter the house. Seal with caulk or expanding foam insulation.
- Check attic insulation. Look for gaps around chimneys; fill with rock wool or high-temperature caulk. Seal gaps where ductwork penetrates the ceiling into the attic.
- Check for gaps and separation in heating ducts in attics and basements. Seal with foil-backed butyl tape, which lasts longer than traditional duct tape.
- Close heating vents and doors to unused rooms.
- Finish interior and exterior painting projects.
- Check caulking around bathtubs, showers and sinks; replace as needed.
- Clean or replace interior and exterior doormats.

- _____
- _____
- _____

GARDENING

- Plant garlic.
- Pinch back tops of Brussels sprouts plants to encourage sprouts to fill out before winter.
- Clean out spent garden beds. Add healthy plant material to compost; dispose of any diseased plants in the trash. Amend soil with lime, manure or compost as needed. Cover beds intended for early spring planting with a thick layer of straw or leaves; plant winter rye on others to hold soil over winter.
- Plant berry bushes and asparagus if not done in September.
- Cut down yellowed asparagus fronds. Put well-rotted manure on beds; mulch with leaves.
- Pot up herbs to bring indoors for winter use.
- Prepare frost covers, plastic tunnels and other winter plant protection. Pay attention to the weather forecast for frost and freeze alerts.
- Set up cold frames if not done yet. Plant with cold-tolerant greens such as spinach, arugula, kale, romaine, endive, 'Winter Marvel' looseleaf lettuce and 'Arctic King' or 'North Pole' butterhead, as well as mache (corn salad), radishes and scallions. Prop open cold frame lids about six inches when outdoor temperatures are above 40 degrees; open entirely when temperature is over 50 degrees.
- Update garden records to plan for next year. What worked well this year? What didn't? What would you like to do differently next year?
- Consider starting new garden beds with the "lasagna" method. Layer cardboard or thick newspapers over the ground where you want the new bed. Removing the sod is not necessary. Throughout the fall, pile layers of manure, compost, leaves and grass clippings on the bed. By spring you'll have a new garden ready for planting!

- _____
- _____
- _____

HARVESTING/PRESERVING

- Pick lettuce, spinach, Swiss chard, Brussels sprouts, parsnips, carrots and other vegetables that are still producing.
- Harvest nuts. Use fresh or store in freezer.
- Pick apples. Eat fresh or make cider, dried apples, applesauce, apple butter and/or baked goods.
- Harvest pumpkins, winter squash and gourds. Cure at 80-85 degrees and about 80% humidity for ten days, then store in a cool (50-55 degrees), dry place.
- Harvest Indian corn and popcorn when stalks have turned brown and kernels are dry and hard. Store in a dry, rodent-proof area.
- Save seeds of non-hybrid vegetable varieties that you like and that performed well in your location. Dry well and store in a dry, dark place.

- _____
- _____
- _____

FORAGING

- **WARNING: Many wild plants or parts of plants can be toxic! Before consuming any wild foods, be sure you know exactly what you are harvesting and how to properly prepare it. Harvest only from clean sites not contaminated with chemical sprays, traffic pollution, animal waste or other toxins. Do not consume large amounts of any wild plant unless you are absolutely sure of its safety.**

- **Before harvesting and consuming wild plant foods, consult an experienced local forager or a reputable foraging field guide such as one of the following:** *A Field Guide to Edible Wild Plants: Eastern and Central North America* **by Lee Allen Petersen or** *The Complete Guide to Edible Wild Plants* **by the Department of the Army.**

- Pick chickweed for salads.
- Pick young plantain leaves; use fresh or cooked.
- Pick young leaves, upper stems and roots of redroot pigweed. Use leaves and stems as fresh or cooked vegetable; cook and mash roots as you would potatoes.
- Harvest dry seeds of dock, bulrush, plantain, redroot pigweed and lamb's quarters. Use as a grain or grind and use for flour and thickener. Redroot seeds can be "popped" like popcorn.
- Harvest the small bean-like seeds from "helicopters" of sugar maples. Eat raw or cooked.
- Harvest remaining chestnuts, shagbark hickory nuts, black walnuts, beechnuts, butternuts and acorns. Use fresh or store in freezer.
- Dig chicory roots; dry and grind for coffee substitute.

- _____
- _____
- _____

LIVESTOCK

- Winterize livestock buildings. Aim only to limit drafts and keep precipitation out; airtight buildings are not healthy for animals or humans.
- Give all livestock housing a thorough cleaning. Sweep down cobwebs, remove old bedding down to bare floors and open windows and doors to let buildings air out on a sunny day. Scrub and disinfect feeders and waterers. Make any needed repairs.

- Check barn doors, gates and windows for safety and security.
- De-clutter and organize tool and equipment storage areas. Sell, trade or give away unneeded items. Consider using any profits to beef up the homestead emergency fund.
- Clean and organize feed storage areas. Storing grain and feed in an old chest freezer helps keep it dry and protects it from rodents, insects and livestock.
- Begin to limit grazing and feed more hay as pastures decline. Consider confining animals to sacrifice lot or paddock to allow pastures to regrow before winter.
- Check fences and gates; repair as needed.
- Check watering equipment. Make improvements as needed.
- Mow and drag pastures and allow them to rest until spring.
- Spread manure on fallow ground.
- Monitor and treat for external parasites as needed.
- Butcher or sell unwanted or unproductive animals.
- Arrange for butchering or sale of young animals that have reached market size. Generally, beef cattle are butchered at the end of their second summer, hogs at about 250 pounds, lambs at 90-100 pounds (about 5-6 months old) and goat kids at 6-8 months. However, animals can be slaughtered at whatever age or weight suits the owner's or buyer's personal preferences and needs.

- _____
- _____
- _____

Bees

- Prepare hives for winter.
- Install mouse guards/entrance reducers if not done in September.
- Feed 2:1 sugar/water solution if needed.

- _____
- _____

Cattle

- Decide whether to raise, butcher or sell calves born this year. Prepare facilities for those you plan to keep.

- _____
- _____

Goats

- Continue fall breeding for open (non-pregnant) does. Estimate due dates on calendar, about 150 days after breeding.

- _____
- _____

Horses

- _____
- _____

Pigs

- Breed sows now for January or February farrowing. Estimate farrowing date on calendar at three months, three weeks and three days after breeding date.
- Prepare farrowing stalls for sows bred in summer. Have a safely installed heat lamp or other source of warmth available for piglets.
- Prepare farrowing kit, including clean towels, iodine and a small cup for navel dipping, elbow-length plastic gloves, OB lubricant, flashlight, nasal aspirator syringe, colostrum and/or milk replacer or goat milk, feeding tube, human baby bottles and nipples.

- Be very careful of your personal safety around farrowing sows and those raising piglets. Even "pet" sows can become dangerously protective of their young.
- Dip navels of newborn piglets. Warm small, listless piglets under heat lamp if needed. Notch or tattoo ears if needed for identification. Be sure all piglets have a chance to nurse. Give iron shots or supplements to piglets at 3-5 days of age if they do not have access to soil in their pen. Clip tips of sharp canine teeth ("needle teeth") if desired. Castrate males at about one week of age.

- _____
- _____

Poultry

- Remove pastured poultry from range before grass declines and weather turns cold.

- _____
- _____

Rabbits

- Examine litters of newborn bunnies; remove dead ones and place excess or rejected ones with foster doe if available.

- _____
- _____

Sheep

- Shear lambs 4-6 weeks before butchering in order to use or sell

pelts as shearling.
- Breed now for March lambing. Estimate due dates on calendar, about 145 days after breeding.
- Market spring-born lambs.

- _____
- _____

PETS

- Trim nails.
- Wash bedding and dishes.

- _____
- _____
- _____

BUILDINGS & GROUNDS

- Chop fallen leaves with lawn mower and add them to compost or use for mulch.
- Plant spring bulbs including daffodils and tulips.
- Dig up and store tender bulbs and plants such as dahlias, canna lilies and gladiolus.
- Take tender plants inside house or greenhouse.
- Finish digging and dividing late-blooming perennials.
- Sprinkle bone meal around irises.
- Plant pansies and flowering kale.
- Finish planting or moving evergreens.

- _____
- _____
- _____

Ponds

- Clean excess fallen leaves out of ponds.
- _____
- _____

MACHINERY & EQUIPMENT

- Check fluid levels in vehicles, taking special note of antifreeze mixture. Put winter tires on, check tire pressure and get vehicles tuned up.
- Lubricate hinges, latches and locks.
- _____
- _____
- _____

ENJOYMENT

- For a healthy campfire treat, roast apples over an open fire. Insert cooking stick into the apple and hold over the hot coals until the peel splits.
- Clean and put away hummingbird feeders.
- _____
- _____
- _____

11

NOVEMBER

Giving Thanks

This month we take the time to be thankful -- for the opportunity to be self-sufficient, for the freedom to live as we choose, and for everything that our homestead provides for us. We can be thankful for the tangible benefits of living on a homestead, such as food, water and shelter, as well as for the intangible benefits like peace of mind and a feeling of harmony with nature. Even if we're not cooking a traditional Thanksgiving feast, we can consider hosting a simple "homestead harvest" dinner or donating to our local food bank. Let's be thankful for all that we have and all that we are able to give from our homesteads!

HOUSEHOLD

- Put fresh batteries in smoke and carbon monoxide detectors. Check expiration dates on fire extinguishers. Replace expired extinguishers or have them professionally recharged.
- If you have a generator and aren't completely familiar with its use, read the manual now and decide how and where you will use the generator in the event of a power outage. Designate a safe area in which to operate the generator during heavy snow or rain. Make sure you have enough fuel to run it for at least three days – or preferably longer.
- Restock emergency household supplies in case of a natural disaster or power outage. Check that you have at least three days' worth of supplies including firewood or heating fuel; flashlights and batteries; nonperishable food including granola bars and any special food you use regularly; water; medicines; battery-operated or hand crank radio; pet food; toilet paper; feminine hygiene items; baby food or formula; diapers; hand sanitizer; a hand can opener and a first aid kit. Make a "bug-out" bag or backpack full of survival supplies for each family member. In case of an evacuation each person can grab his or her bag and go.
- Store important documents in a fireproof safe near your home evacuation/emergency supply kits.
- Post a list of emergency phone numbers by your landline phone and program them into your cell phones. Include the fire and police departments, poison control center, hospital, doctors, neighbors, veterinarian and farrier.
- Post written directions to your homestead in the kitchen or near a landline phone for use when calling for help in an emergency.
- Clean, cover or bring in grill, planters, bird baths, outdoor furniture and toys.
- Cover drafty windows with heavy draperies, quilts or clear plastic.
- Drain and shut off outdoor hoses and faucets. Turn off or insulate water lines to unheated buildings.
- Vacuum dryer vents, ducts and lint traps to remove lint buildup.
- Vacuum refrigerator and freezer coils, clean gaskets and lubricate hinges. Unplug appliances before vacuuming coils.
- Test pre-2006 GFCI breakers in kitchen and bathrooms with a

circuit tester, available in home centers and hardware stores. GFCIs made after 2006 will simply stop providing electricity when they no longer function.
- Sharpen kitchen and hunting knives.
- Clean and organize pantry food storage areas; discard outdated items.
- Change or clean household water filter according to manufacturer's instructions.
- Check furnace filter; clean or change as needed.
- Buy non-toxic sidewalk de-icer. In a pinch you can use cat litter, sand, sawdust or wood shavings to provide traction on icy surfaces. Be careful not to track gritty substances onto hardwood floors.

- _____
- _____
- _____

GARDENING

- Bring in garden tools. Clean metal surfaces, sharpen cutting edges and spray with WD-40 or wipe with motor oil. Clean handles. Wipe bare wood with linseed oil; touch up or repaint painted handles.
- Organize tool storage area and store tools for winter.
- Scrub empty pots and planters with mild bleach solution, dry thoroughly and store for winter.
- Drain and put away garden hoses.
- Blow water out of drip lines with air compressor.
- Turn compost. If bin is full, cover with 2" of soil and water well, then start a new bin.
- Continue to clean out garden beds as they finish producing. Remove dead plant material. Add healthy material to compost; dispose of diseased plants in the trash. Add lime, manure, compost or other amendments to garden beds as needed. Cover beds with mulch or plant winter rye for "green manure" if desired.

- Prepare early spring planting beds now. Loosen soil, add amendments as needed and mulch with a thick layer of straw or leaves. Lay branches over the mulch to keep it from blowing away.
- Put a thick layer (about a foot) of straw or leaves on crops that you wish to leave in the ground for late fall harvest, such as parsnips, carrots and leeks. Do this after a few frosts but before the ground freezes.
- Cut back to ground level all fall-bearing raspberry canes that have fruited. Dispose of all cuttings in the trash to reduce the spread of pests and disease organisms.
- Clean up any remaining fallen fruit around apple and pear trees.
- Make shopping list of needed tools and supplies.

- _____
- _____
- _____

HARVESTING/PRESERVING

- Harvest Brussels sprouts, spinach, Swiss chard, leeks, kale, parsnips, Indian corn, pumpkins, broccoli and turnips.
- Continue to harvest salad greens.
- Pick apples.
- Monitor stored vegetables and fruits for mold, rot or rodent damage.

- _____
- _____
- _____

FORAGING

- **WARNING: Many wild plants or parts of plants can be toxic! Before consuming any wild foods, be sure you know exactly what you are harvesting and how to properly prepare it. Harvest only from clean sites not contaminated with chemical sprays, traffic pollution, animal waste or other toxins. Do not consume large amounts of any wild plant unless you are absolutely sure of its safety.**
- **Before harvesting and consuming wild plant foods, consult an experienced local forager or a reputable foraging field guide such as one of the following:** *A Field Guide to Edible Wild Plants: Eastern and Central North America* **by Lee Allen Petersen or** *The Complete Guide to Edible Wild Plants* **by the Department of the Army.**

- Pick chickweed for salads.
- Pick young leaves, upper stems and roots of redroot pigweed. Use leaves and stems as fresh or cooked vegetable; cook and mash roots as you would potatoes.
- Harvest dry seeds of dock, bulrush, plantain, redroot pigweed and lamb's quarters. Use as a grain or grind and use for flour and thickener. Redroot seeds can be "popped" like popcorn.
- Pick rose hips for herbal tea.
- Harvest remaining hickory nuts, black walnuts, beechnuts, butternuts and acorns. Use fresh or store in freezer.
- Dig chicory roots; dry and grind for coffee substitute.
- Dig daylily tubers; use as you would potatoes.
- Dig Jerusalem artichoke tubers. Scrub thoroughly; eat raw or cooked.

- _____
- _____
- _____

LIVESTOCK

- As pastures decline, monitor animals' body condition and give supplemental feed if needed. Refer to one of the many livestock body condition charts available online or through your county Extension agent.
- For best pasture management, remove grazing animals from declining pastures. Begin feeding hay in dry lots or move animals to reserved pastures.
- Mow any remaining tall weeds in pastures.
- Check fences and gates; repair as needed.
- Be sure that pasture-kept animals have adequate shelter from wind and precipitation.
- Test for internal parasites; treat as needed. Consult your veterinarian for recommendations regarding deworming routines and medications, especially for animals providing milk and meat for human consumption.
- Prepare a disaster plan for your livestock. Photograph yourself with the animals and keep these photos in your home evacuation kit. Decide what you will do with the animals in case of disaster and keep written instructions in an accessible place. If you have a stock trailer and plan to take animals with you, make sure they will load onto the trailer willingly and that you can take enough hay, feed and water to last several days. Keep an animal first aid kit in the trailer or truck.
- Identify at least two neighbors or friends who could care for your animals if you were sick or had to be away from home for several days. Walk them through the daily routine and post written instructions in the barn, along with your cell phone numbers and contact information for the veterinarian, farrier and feed supplier.
- Post written directions to your homestead in the barn for use when calling for help in an emergency.
- Arrange for butchering or sale of young animals that have reached market size. Generally, beef cattle are butchered at the end of their second summer, hogs at about 250 pounds, lambs at 90-100 pounds (about 5-6 months old) and goat kids at 6-8 months. However, animals can be slaughtered at whatever age or weight suits the owner's or buyer's personal preferences and needs.

- _____
- _____
- _____

Bees

- Clean up the area around the hives. Cut back weeds and rake leaves.
- Check that hives are waterproof, windproof and protected from livestock and predators.
- Be sure each hive has an upper entrance in case of deep snow.
- Be sure hives have adequate ventilation to allow moisture to evaporate.
- Feed bees 2:1 sugar/water syrup if honey supply is low.
- Continue reducing entrances.

- _____
- _____

Cattle

- _____
- _____

Goats

- Check hooves; trim if needed.
- Breed now for April kidding. Estimate due dates on calendar, about 150 days after breeding.

- _____
- _____

118 THE COMPLETE HOMESTEAD PLANNER

Horses

- Trim hooves.
- Cut back on grain as work decreases.

- _____
- _____

Pigs

- Provide adequate shelter and bedding as temperatures dip.
- Breed sows now for February or March farrowing. Estimate farrowing date on calendar at three months, three weeks and three days after breeding date.
- Purchase or raise fall feeder pigs for Easter hams.
- Butcher hogs when they reach about 250 pounds.
- Prepare farrowing stalls for sows bred in summer. Have a safely installed heat lamp or other source of warmth available for piglets.
- Prepare farrowing kit, including clean towels, iodine and a small cup for navel dipping, elbow-length plastic gloves, OB lubricant, flashlight, nasal aspirator syringe, colostrum and/or milk replacer or goat milk, feeding tube, human baby bottles and nipples.
- Be very careful of your personal safety around farrowing sows and those raising piglets. Even "pet" sows can become dangerously protective of their young.
- Dip navels of newborn piglets. Warm small, listless piglets under heat lamp if needed. Notch or tattoo ears if desired for identification. Be sure all piglets have a chance to nurse. Give iron shots or supplements to piglets at 3-5 days of age if they do not have access to soil in their pen. Clip tips of sharp canine teeth ("needle teeth") if desired. Castrate males at about one week of age.

- _____
- _____

Poultry

- Refresh nest material.
- Scrub and air-dry feeders and waterers.
- Select and harvest your Thanksgiving turkey.

- _____
- _____

Rabbits

- Remove young from doe's hutch at six weeks of age.
- Clean cages, hutches and nest boxes. Scrub feeders and waterers.
- Add bedding for winter warmth.
- Make sure rabbits are out of drafts and that they have hay and fresh unfrozen water available.

- _____
- _____

Sheep

- Breed now for April lambing. Estimate due date on calendar, about 145 days after breeding.
- Check hooves; trim if needed.
- Keep sheep areas clean and dry to maintain cleanliness of wool.

- _____
- _____

PETS

- Wash bedding and dishes.
- Wash litter boxes and allow to dry in the sun.
- Trim nails.
- _____
- _____

BUILDINGS AND GROUNDS

- Clean leaves and debris out of gutters.
- Insulate springhouses and wellheads if necessary.
- Dig hole now if you plan to plant a live Christmas tree. Fill hole with leaves and cover with boughs or branches until planting time.
- Begin clearing trees and brush. Pile cut brush for wildlife cover, use as kindling for bonfires or shred for mulch.
- Once leaves have fallen, find and destroy wasp nests hanging in trees.
- Finish planting trees and bulbs. Water new trees well until ground freezes.
- Consider leaving dead and dried perennials standing over the winter, as long as they are not diseased. Dried stalks provide cover for beneficial insects, as well as naturally collecting leaves and snow to insulate the living roots underneath. If you don't like how they look, cut plants down and compost healthy material; dispose of diseased stems and leaves in the trash.
- Remove broken or diseased branches from trees. Check for weak or damaged branches that may fall during heavy snow loads.
- Mulch hollies, rhododendrons and azaleas with fallen leaves, pine needles or shredded pine bark to protect roots and conserve moisture.
- To propagate forsythia and privet for spring planting, take cuttings and store over winter in damp sand in cellar.
- Clean up fallen leaves. Chop with mower and add to compost or use as mulch.

NOVEMBER 121

- Mow wildflower meadows to scatter seed.
- Put reflectors or other markers along driveway to guide winter snow plowing.

- _____
- _____
- _____

Ponds

- Remove fallen leaves to reduce tannin buildup.
- Remove delicate fish and tender plants from small ponds.
- Drain small artificial ponds and fill with leaves or branches.
- Clean pump systems.
- Check that pond overflow pipes are clear of leaves and other debris.

- _____
- _____

MACHINERY & EQUIPMENT

- Check vehicle fluid levels and tire pressure according to owners' manuals.
- Put winter tires on vehicles if not done yet.
- Restock winter emergency kit in cars and trucks. Include flashlights with fresh batteries, a small shovel, duct tape, bungee cords, tow rope or chain, jack, lug wrench, socket wrenches, open-end wrenches, tire iron, screwdrivers, pliers, fire extinguisher, jumper cables, clean rags, hand cleaner, a pencil and notepad, first aid kit, small amount of cash, ice scraper, hand warmers, a blanket, bag of cat litter for traction, emergency LED beacons or warning triangles, small tire pump, spray-in tire sealant, sleeping bag or blankets, extra hat, scarf, gloves,

reflective vest, boots and snow pants. Store several bottles of water, energy bars, nuts and/or dried fruit in a small cooler in the trunk.
- Keep a plastic basket in trunk with small well-marked bottles of replacement fluids.
- Make sure spare tire fits vehicle and is properly inflated.
- Keep the gas tank as full as possible during the winter months.
- Before storing lawn mower for the winter, siphon the gas from the tank or use the mower until the gas is gone. If the tank is full and you can't readily empty it, add a fuel stabilizer. Change the oil. Clean the caked grass and gunk off the underside of the mower deck. Clean or replace air filter, charge the battery, change worn-out spark plugs and sharpen the blade. Store mower in a dry place.
- Prepare snow plowing equipment.

- _____
- _____
- _____

ENJOYMENT

- Begin feeding wild birds.
- Host a harvest dinner or potluck party for friends or neighbors.
- Look forward to the coming winter knowing that you are well prepared!

- _____
- _____
- _____

12

DECEMBER

Resting and Reviewing

Another year has come full circle and winter is upon us. Now that the busy season of outdoor work has come to a close, it's time to sit back and reflect on the year's successful and not-so-successful efforts. What worked well this year? What didn't? What can you add, drop or change to make next year even better? Keep a notebook handy throughout the month to jot down ideas, plans and goals for the homestead. By the end of the month, you'll be ready to make decisions for next year's growing season.

HOUSEHOLD

- Check heating fuel supplies and order more if necessary.
- Cut and split firewood.
- Check furnace filter; clean or change as needed.
- Test smoke and CO detectors according to manufacturers' instructions.
- Clean light fixtures to allow maximum brightness on long winter nights.
- Set a holiday budget and stick to it.

- _____
- _____
- _____

GARDENING

- Stack straw bales around cold frames and pile loose straw on top if temperatures are extremely cold. Once the harvest is over for the season, clean out remaining plant material. Make any needed repairs to the cold frames and put them away if they are not permanent fixtures in the garden.
- Clean and put away tools if not done in November. Clean metal surfaces, sharpen cutting edges and spray metal parts with WD-40 or wipe with motor oil. Clean handles. Wipe bare wood with linseed oil; touch up or repaint painted handles.
- Scrub empty clay and plastic flowerpots, soak in mild bleach solution, rinse well and store for winter.
- Begin to plan next year's garden.
- Consider forming or joining a seed-buyers' co-op. Plan to get a group of gardening friends together in January to order seeds together from one catalog to save money through bulk discounts and combined shipping. Everyone should bring a list of what they'd like to order, and then combine their orders into one. Designate a leader to record and submit the final order. Plan to meet again to distribute the seeds when they arrive. For the second

meeting, everyone should bring jars, plastic bags or envelopes for their share of the seeds, as well as markers to label the packages. To make the meetings even more enjoyable, have everyone bring a covered dish or dessert to share.

- _____
- _____
- _____

HARVESTING/PRESERVING

- If the weather is mild, you may still be able to harvest Swiss chard, spinach, romaine, Brussels sprouts, kale and root vegetables.

- _____
- _____
- _____

FORAGING

- **WARNING: Many wild plants or parts of plants can be toxic! Before consuming any wild foods, be sure you know exactly what you are harvesting and how to properly prepare it. Harvest only from clean sites not contaminated with chemical sprays, traffic pollution, animal waste or other toxins. Do not consume large amounts of any wild plant unless you are absolutely sure of its safety.**

- **Before harvesting and consuming wild plant foods, consult an experienced local forager or a reputable foraging field guide such as one of the following:** *A Field Guide to Edible Wild Plants: Eastern and Central North America* **by Lee Allen Petersen or** *The Complete Guide to Edible Wild Plants* **by the Department of the Army.**

- Pick rose hips; dry for tea.
- Dig chicory roots; dry and grind for coffee substitute.
- Dig Jerusalem artichoke tubers. Scrub thoroughly; eat raw or cooked.
- Dig daylily tubers; use as you would potatoes.

- _____
- _____
- _____

LIVESTOCK

- Feed extra hay in very cold or windy weather. The metabolic processes of digesting roughage will help to keep animals warmer.
- Make sure all animals have access to unfrozen water. Break ice several times a day if necessary. Try floats or insulated buckets to help keep water liquid. For a DIY alternative, stack two tires, stuff cavities with straw or hay and nestle a water bucket in the center. Electric buckets and tank heaters should be used only with caution and frequent monitoring. Cords and tank heaters must be inaccessible to animals and should be grounded with a GFCI outlet to prevent electric shock.
- Make sure pastured animals have adequate shelter from wind and precipitation.
- Check pastures and paddocks for icy areas. Keep animals away from ice or scatter sand or sawdust over it for traction.
- Check fences and gates; make repairs as needed.
- Check supplies of feed and bedding.

Bees

- Check honey supply in each hive and feed fondant or sugar brick if needed.

Cattle

- Keep udders, teats and bedding dry to prevent frostbite in below-zero temperatures.
- Cull unproductive or unhealthy cows.

Goats

- Get remaining open does bred.
- Cull unproductive or unhealthy does.
- Sell kids.

Horses

- Put snow pads under shoes of shod horses. Watch hooves for snowball buildup.
- Clean and repair tack.
- _____
- _____

Pigs

- Provide extra bedding in frigid temperatures.
- _____
- _____

Poultry

- Add litter to floor of coop if needed.
- Cull unproductive or unhealthy birds.
- Gather eggs at least twice a day if temperatures are below freezing.
- _____
- _____

Rabbits

- Check twice a day to be sure rabbits have unfrozen water to drink.
- Clip nails on breeding stock.
- Separate male and female young before they are three months old.
- If raising rabbits for meat, butcher between eight weeks and eight months of age.

- _____
- _____

Sheep

- Sell spring-born lambs for the holiday market.
- Breed now for May lambing. Estimate due date on calendar, about 145 days after breeding.

- _____
- _____

PETS

- Trim nails.
- Keep holiday plants out of reach of pets.
- Wash pet bedding.

- _____
- _____
- _____

BUILDINGS & GROUNDS

- Mulch perennials and bulbs if not done in November.
- Put chew-proof guards around small tree trunks if rabbits and rodents are gnawing on them. Keep mulch several inches away from base of trunk to reduce rodent hiding places.
- Finish cleaning up leaves if not done in November. Use them for mulch or add to compost.
- Water rhododendrons before ground freezes.
- Mulch newly planted trees and shrubs after ground freezes.
- Apply deer repellent or barriers to vulnerable evergreens and

roses. One method is to encircle the plant with a ring of stakes or T-posts driven into the ground about a foot away from the plant, then wrap burlap or chicken wire around the outside perimeter of the ring.
- During heavy snowfalls, periodically shake snow from evergreens, but do not attempt to remove ice. Doing so may damage the plant.

- _____
- _____
- _____

MACHINERY & EQUIPMENT

- Check vehicle fluids and tire pressure according to owners' manuals.
- Wash vehicles periodically if roads are salted.

- _____
- _____
- _____

ENJOYMENT

- String fresh cranberries and popped popcorn to put around outdoor evergreens for the birds.
- Put up suet feeders for birds or simply smear suet or peanut butter directly onto tree bark.
- To start the year off right, celebrate New Year's Eve the way you've always wanted to, whether it's by attending your local public celebration, hosting an all-out party for friends or simply spending a quiet evening reflecting on the past year and dreaming of the year to come.

DECEMBER **131**

Made in the USA
Lexington, KY
31 January 2015